USED MATH FOR THE FIRST TWO
YEARS OF COLLEGE SCIENCE

USED MATH FOR THE FIRST TWO YEARS OF COLLEGE SCIENCE

Clifford E. Swartz

*State University of New York
at Stony Brook*

with the assistance of

*Paul A. Swartz and
B. Katherine Swartz*

American Association of Physics Teachers
5112 Berwyn Road
College Park, MD 20740-4100
U.S.A.

ISBN # 0-917853-50-4

CONTENTS

Appendices, 227

Index, 259

PREFACE

This is not a mathematics text. We assume that the user has already had a formal introduction to most of the topics covered here, or is now studying them in a math class. It is possible, however, that attention to the intrinsic beauty of the formal structure has left little time for mundane applications. It turns out that simple mathematics is almost unreasonably successful in modeling phenomena of the real world. There is some point in learning to manipulate the model, even though its inner workings are not of personal concern. To be sure, somebody ought to study the model for its own sake if only to assure us that it does not have inner weaknesses or inconsistencies. But for many of our purposes, we need only know the range of validity of the model and how to work it.

In this book, which is part reference and part reminder, we are concerned with how to use math. We concentrate on those features that are most needed in the first two years of college science courses. That range is not rigorously defined, of course. A sophomore physics major at M.I.T. or Cal. Tech. must use differential equations routinely, while a general science major at some other place may still be troubled by logarithms. It is possible that even the Tech student has never really understood certain things about simple math. What, for instance, is natural about the natural logs? We have tried to cover a broad range of topics—all the things that a science student might want to know about math but has never dared ask.

The sections of the book are not sequential. Because of our assumption about the user's prior exposure to math, we have felt free to make use of any topic or notation to help explain any other topic. One of our purposes is to demonstrate such cross-links. The book itself is extensively cross indexed and we have attempted to make the index at the back of the book as complete as possible.

The idea of writing a book like this arose many years ago when I discovered that a whole class of physics students, taking calculus concurrently, knew that $\sin \theta \simeq \theta$ for small θ. None of them, however, knew that the approximation is meaningful only if you use radians. When anyone learns new math he usually gleans only a superficial understanding. For math to be understood, it must be old and used.

I am delighted that the American Association of Physics Teachers has decided to re-issue *Used Math*. The original edition was published in 1973. Some changes have taken place since then. I corrected a few typos, and decided that it was no longer necessary to print pages of log and trig tables. However, I have added some extra comments about error analysis for students, a perennial subject of controversy at every level. The two M.I.T. students who helped in the first edition and were acknowledged on the title page, now have Ph.D.'s and are productive professionals, one an economist and the other a physicist.

I've used the book for reference a lot in the last 20 years. I hope that a new generation of teachers and students will also find it useful.

C.E.S.

USED MATH FOR THE FIRST TWO YEARS OF COLLEGE SCIENCE

CHAPTER 1 REPORTING AND ANALYZING
UNCERTAINTY

When a value is not known precisely, the amount of uncertainty is usually called "error." This has given the whole business of uncertainty analysis a bad name, because in common usage "error" implies sloppiness, very likely caused by sinfulness. (She saw the error of her ways.)

In the language of technology, this section deals with the analysis of errors—how to judge their magnitude, how to describe them in conventional ways, and how to take them into account in calculating numerical values based on a number of individual measurements. As we shall emphasize, *error* represents *uncertainty* and has nothing to do with mistakes or sloppiness. Indeed, reducing the amount of error in a measurement beyond the immediate need is usually a mistake and a sign of poor judgment.

We categorize the reporting and handling of uncertainty in four stages, each involving a particular degree of precision. The first is concerned only with order of magnitude of a number. Consider this to be a zeroth approximation. Next are the conventions regulating the use of significant figures, a first approximation of limited usefulness. The second approximation deals with the maximum and minimum range of measured quantities. The rules of manipulation of such error limits are simple, and this system should be the one most often used by students of introductory science. Finally, there is a third approximation to error citation and analysis, involving rules derived from probability and statistics. This system is frequently misunderstood and misapplied. Its use is justified only when the primitive data fulfill certain requirements of quantity, distribution, and probability.

1.1 HOW TO WRITE NUMBERS—THE EXPONENTIAL NOTATION. The exponential notation is just a means of indicating and keeping track of the decimal point position in a number. However, use of the method provides a powerful aid in doing arithmetic problems or setting them up for slide-rule solution. The notation is particularly important in dealing with very large or very small numbers. Because students often meet exponential notation for the first time in science classes, the system is sometimes called "scientific notation." Fortunately, this terminology is unknown in the real world outside the classroom.

0.001	0.01	0.1	1	10	100	1000	10,000
10^{-3}	10^{-2}	10^{-1}	10^{0}	10^{1}	10^{2}	10^{3}	10^{4}

Note that the 2 in 10^2 can be thought of as either the number of zeros in 100 or the number of times that 10 is multiplied by itself—10 squared. Negative exponentials indicate the reciprocal power of the number: $10^{-2} = 1/10^2 = 1/100 = 0.01$.

To multiply the powers of 10, add the exponentials: $100 \times 1000 = 10^2 \times 10^3 = 100,000 = 10^5$. $0.1 \times 0.001 \times 100 = 10^{-1} \times 10^{-3} \times 10^2 = 0.01 = 10^{-2}$. This property justifies the assignment $10^0 = 1$. We must have this equality since if $10 \times 0.1 = 1$, then $10^1 \times 10^{-1} = 10^{1-1} = 10^0 = 1$.

Here are some examples of numbers written in exponential notation:

$$15 = 1.5 \times 10^1 \qquad 5,380,000 = 5.38 \times 10^6 \qquad 9 = 9 \times 10^0$$
$$90.0 = 9.00 \times 10^1 \qquad 90 = 9 \times 10^1 \qquad 0.0032 = 3.2 \times 10^{-3}$$

Notice the difference between writing 90.0 and 90. By convention the first number indicates that the value is known to be between 89.95 and 90.05. The extra zero is significant and must be retained. The form of the second number may be ambiguous. The value may be known to be between 89.5 and 90.5 or perhaps only between 85 and 95. The exponential notation provides a way of avoiding the ambiguity. To indicate the former value, we would write 9.0×10^1; to indicate the latter value, we would write 9×10^1. (For rules concerning significant figures, see p. 6.)

An example of how to set up a multiplication and division problem is given below. This procedure should be used, especially if the problem is going to be solved with slide rule or logs.

$$\frac{5832 \times 0.051}{68,000 \times 32} = \frac{5.832 \times 10^3 \times 5.1 \times 10^{-2}}{6.8 \times 10^4 \times 3.2 \times 10^1} = \frac{5.8 \times 5.1}{6.8 \times 3.2} \frac{10^1}{10^5}$$
$$= 1.4 \times 10^{-4}$$

Using this technique, a problem like this can quickly be solved to one or two significant figures by inspection, and the placement of the decimal point determined. A slide rule, logs, or pencil-and-paper calculation can then be used to add whatever other significant figures are needed and are justified. For instance, the third step above sets up the problem so that it can be solved approximately by looking at it. The numerator product must be about 30 (6×5); the denominator product must be about 20 (7×3).

Therefore, the answer must be about 1.5×10^{-4}. A slide-rule check yields 1.4. If the slide rule had not read close to 1.5, we would have suspected that a mistake had been made.

Notice that, in this example, no more significant figures are retained in any one of the numbers to be multiplied than belong to the number having the least significant figures. Since 32 and 0.051 have only two significant figures, 5832 is rounded off to 5.8×10^3. When using a slide rule, however, it is usually just as easy to set each number to three significant figures.

Raising a number to some power, or taking a root, is easy to do with exponential notation. For example, $(10^2)^3 = 10^2 \times 10^2 \times 10^2 = 10^6 = 10^{2 \times 3}$. When raising to a power, multiply the exponents. When taking a root, divide the exponents: $\sqrt{10^6} = 10^{6/2} = 10^3$. Here are some more complicated examples.

$$(5800)^2 = (5.8 \times 10^3)^2 = (5.8)^2 \times 10^6 \qquad (\approx 36 \times 10^6) = 34 \times 10^6$$

The symbol $\approx$ means "approximately equal." The parenthetical value is the approximation obtained just by looking at the number.

$$\sqrt{0.47} = \sqrt{47 \times 10^{-2}} = \sqrt{47} \times 10^{-1} \qquad (\approx 7 \times 10^{-1}) = 0.69$$

(Arrange the exponent of 10 so that it is easily divisible by the root. For instance, it would not be sensible to write $\sqrt{0.47} = \sqrt{4.7 \times 10^{-1}} = \sqrt{4.7} \times 10^{-1/2}$, etc.)

$$\sqrt[3]{6 \times 10^{23}} = \sqrt[3]{600 \times 10^{21}} = \sqrt[3]{600} \times 10^7 \approx 8 \times 10^7.$$

SAMPLES

1. $$\frac{14,000}{383 \times 1.0\%} = \frac{14 \times 10^3}{3.83 \times 10^2 \times 1.0 \times 10^{-2}} = \frac{14}{3.83} \frac{10^3}{10^0}$$
$$(\approx 3\tfrac{1}{2} \times 10^3) = 3.7 \times 10^3$$

By leaving 14,000 as 14×10^3 instead of 1.4×10^4, we made the numerator larger than the denominator, and so ended with a quotient greater than 1—an unnecessary but convenient trick. The parenthetical value of $\approx 3\tfrac{1}{2} \times 10^3$ is the approximation obtained just by looking at the fraction.

2. $$\frac{0.084 \times 86}{9.2 \times 74} = \frac{8.4 \times 10^{-2} \times 8.6 \times 10^1}{9.2 \times 7.4 \times 10^1} \qquad \left(\approx \frac{70}{68} \times 10^{-2}\right) = 1.1 \times 10^{-2}$$

3. $$\frac{4623 \times 98 \times 41}{0.0062 \times 122} \qquad (2.5 \times 10^7)$$

4. $$\frac{18}{6 \times 10^{23}} \qquad (3 \times 10^{-23})$$

5. $$\frac{10,001 \times 0.08}{143 \times 9.87} \qquad (0.6)$$

1.2 FERMI QUESTIONS—ORDER OF MAGNITUDE. Fermi questions were named after Enrico Fermi, the great physicist who contributed to both experiments and theory concerned with atomic nuclei and fundamental particles. (He was a master at computing approximations to answers when it seemed that no information was available.) The point of such questions is that reasonable assumptions linked with simple calculations can often narrow down the range of values within which an answer must lie. The *order of magnitude* refers to the power of 10 of the number that fits the value. To increase an order of magnitude means to increase by a factor of 10. Very often an order of magnitude calculation is all that the interest in a problem justifies. Even if more precision is needed, an order of magnitude calculation done first may indicate whether or not it is worthwhile to pursue the problem, and sometimes may indicate how the next approximation to the required value can best be obtained.

Here are some Fermi questions:

A. How many golf balls will fit in a suitcase? Assume that the suitcase is $30'' \times 8'' \times 24''$. The volume is $30'' \times 8'' \times (100/4)'' \approx 6 \times 10^3 \, in^3$. Assume that each golf ball is a sphere 1 in. in diameter. The volume of the ball is a little less than $1 \, in^3$. The order of magnitude of the number of golf balls that will fit in a suitcase is 10^4.

This question could not be taken too seriously unless it were asked by a traveling golf ball salesman. Since the size of the suitcase is not specified, there is no point going to the closet to measure a real one. Imagine a reasonable size. Nor is there any point in worrying about whether or not the balls are close packed as nested spheres; the packing factor could not be greater than 1.5. Surely the diameter of a golf ball is closer to 1 in. than 2 in. Doubling the diameter would increase the ball volume by a factor of 8. That would reduce the number of balls in the suitcase by an order of magnitude. Consider the reasonableness of the order of magnitude answer. Surely the number of golf balls that would fit in a standard suitcase must be greater than 1000 and less than 100,000. Our answer is good within a factor of 10.

B. How many piano tuners are there in New York City? Assume 10^7 people in New York and 2×10^6 families. Assume 1 piano for every 5 families; therefore, 4×10^5 pianos. Assume each piano tuned once every 2 years; therefore, 2×10^5 pianos tuned each year. Assume each tuner tunes 2 a day for 250 days a year. (At $10 per tuning, he barely makes a living—a factor of 2 could make a big difference to the tuner.) Therefore, $(20 \times 10^4$ tunings per year)/(500 tunings per year per tuner) $= 400$ tuners.

It is unlikely that New York City has less than 40 piano tuners or more than 4000. If you do not like the assumptions made, choose your own reasonable guesses and see if your answer is not of the same order of magnitude.

REPORTING AND ANALYZING UNCERTAINTY

Note that sometimes in these calculations one significant figure is carried along. The extent to which you do this depends on the problem and your style; rules would be cumbersome and probably useless. For instance, whether 400 is of order of magnitude 10^2 or 10^3 is a silly question, because a reasonable answer depends on the meaning of the number and the way it is going to be used. When in doubt, carry an extra figure along. Note, incidentally, that a factor of 2 in one of the assumptions makes a big difference to the piano tuners but not to the final result of our order of magnitude calculation.

c. How many cells are there in a human body? Assume that the average cell diameter is 10 microns $(\mu) = 10^{-5}$ meter (m). Then, volume $= 10^{-15}$ m^3. The order of magnitude of human volume is 10^{-1} m^3. Therefore, there are 10^{14} cells in a human body.

This question illustrates again how some information can be obtained out of very little definite knowledge. Living cells come in a great range of sizes. However, they can all be seen with an ordinary light microscope and therefore must have a diameter larger than the wavelength of light. They can scarcely be seen with the unaided eye and so must have a diameter smaller than 0.1 millimeter (mm). We assumed that the diameter was the geometric mean between these values. (The geometric mean of A and B is $\sqrt{A \times B}$. In this case, it is $\sqrt{(10^{-6})(10^{-4})} = 10^{-5}$. The arithmetic mean would be practically the same as 10^{-4}.) Notice that for these calculations it makes no sense to differentiate between the volume of a sphere and that of a cube. The volume of the human body could be estimated by assuming a reasonable height, width, and thickness of a column that is human size. An alternative method requires knowing that 1 liter of water has a mass of 1 kilogram (kg) and a volume of 1×10^{-3} m^3. A cubic meter of water (or flesh) would therefore have a mass of 1000 kg and would weigh about a ton (1 kg weighs 2.2 pounds; therefore 1000 kg weighs 2200 pounds or 1.1 tons). The assumed volume for the body was $\frac{1}{10}$ m^3.

SAMPLES

1. How many hairs are on a human head? (Assume that the spacing between hairs is 1 mm. Then there are 10/centimeter (cm) along a line or 100/cm^2. We figure 2×10^4 hairs on a human head. Do your assumptions lead to results of the same order of magnitude?)

2. How many individual frames of film are needed for a feature length film? (We get 1.5×10^5.)

3. What is the ratio of spacing between gas molecules to molecular diameter in a gas at standard temperature and pressure? [A mole (6×10^{23}) of gas molecules at STP occupies 22.4 liters. A molecular diameter might be 2×10^{-8} cm. Compare available volume per molecule with the volume of a molecule. We get (spacing between molecules)/(molecular diameter) $= 10$.]

4. There are how many seconds in a year? [We get 3×10^7 seconds (s)/year, or 10 megapi s/year ($\pi \times 10^7$ s/year).]

5. If your life earnings were to be doled out to you at a rate of so much per hour for every hour of your life, how much is your time worth? [Perhaps, \$0.50/hour (hr)?]

6. What is the weight of the solid garbage thrown away by American families each year? (We figure about 10^8 tons.)

7. How many molecules are in a standard classroom? ($\sim 10^{28}$)

1.3 SIGNIFICANT FIGURES—A FIRST APPROXIMATION TO ERROR ANALY-SIS. The number of significant figures in a numerical value is a first approximation to showing the limits within which the value is known. There are more precise ways of indicating the error limits; the next approximation is shown on p. 8.

The generally accepted conventions for writing significant figures are summarized in A.

A. 1. When we say that a quantity has the value 3, we mean—by convention—that the value could actually be anywhere between 2.5 and 3.5.

$$2.5 < \underline{3} < 3.5$$

However, if we say that the value is 3.0, then we mean that it lies between 2.95 and 3.05

$$2.95 < \underline{3.0} < 3.05$$

2. Note the ambiguity of a number such as $\underline{300}$. Does it imply $\underline{250 < 300 < 350}$ or $\underline{299.5 < 300 < 300.5}$? A superior method of writing such a number is to use power of 10 notation. In that form the number is $\underline{3 \times 10^2}$ or $\underline{3.00 \times 10^2}$, depending on which precision is intended.

3. 0.000,01 has one significant figure. (It could be written 1×10^{-5}.)
1.000 has four significant figures.
1.000,01 has six significant figures.

Rules for the proper use of significant figures in addition or subtraction are illustrated in B.

B.

5.2	6.843	6.843	6.843	500	5.00×10^2
+3.1	+1.2	+1	+0.001	−4	−4
8.3	8.0	8	6.844	500	4.96×10^2

In addition or subtraction, the sum or difference has significant figures only in the decimal places where *both* of the original numbers had significant

REPORTING AND ANALYZING UNCERTAINTY

figures. This does not mean that the sum cannot have more significant figures than one of the original numbers. In the examples in B, note that 0.001 has only one significant figure, but the sum properly has four. It is the decimal *place* of the significant figure that is important in addition and subtraction. In the final examples in B there is another example of the ambiguity of final zeros. If you estimate that there are 500 students in a lecture, implying a number between 450 and 550, your estimate is not changed if 4 people leave. On the other hand, if you draw out $500 from the bank and spend $4, you have $496 left.

Rules for the proper use of significant figures in multiplication or division are illustrated in C.

C.
$$\begin{array}{r} 5.2 \\ \times 3.1 \\ \hline 52 \\ 156 \\ \hline 16.12 \end{array} = 16 \qquad 5.243 \times 3.1 = 16 \qquad \frac{37}{9} = 4$$

$$\frac{156}{16.12} = \underline{16} \qquad 5.243 \times 0.0031 = 0.016 \qquad \frac{37}{9.1} = 4.1$$

In multiplication and division, illustrated in C, the product or quotient cannot have more significant figures than there are in the least accurately known of the original numbers. Consider the first example: the product might be as large as $(5.25)(3.15) = 16.5375$ or as small as $(5.15)(3.05) = 15.7075$. The rule for significant figures in multiplication is evidently justified in this case. Usually, during multiplication or division, an extra significant figure is carried along, and the final answer is then rounded off appropriately.

SAMPLES

1. $2.000 + 0.01 = 2.01$ Justify the answers in 1 and 2 by calculating the extremes in the sum and product which could be justified by the extremes in the original numbers.

2. $2.000 \times 0.01 = 0.02$

3. $\dfrac{4832 \times 0.165}{15 \times 264} = \qquad (0.20)$

4. What is the volume of a piece of chalk that is 10.5 cm long and has a diameter of 1 cm? (volume $= 8$ cm^3)

5. What is the volume of a rectangular box with length 3.025 cm, width 2.5 cm, and height 2 cm? (By taking extremes of the significant figures, show why it would be unreasonable to cite the answer with only one significant figure. A good answer would be volume $= 15$ cm^3. This variation of the rule applies when the first digit is 1. The special rule and the indeterminancy point up the need for a more precise way of specifying and dealing with error limits on numbers.)

1.4 ABSOLUTE AND PERCENTAGE ERRORS—A SECOND APPROXIMATION TO ERROR ANALYSIS.

This second approximation to error statement and analysis is based on maximum pessimism. The absolute error, the ± 2 cm in the first example in A, defines the maximum excursion of the main value. The implication is that all the lengths measured and the true length fall between 95 and 99 cm.

A. Example of absolute error: 97 ± 2 cm $\qquad 12 \pm 2$ cm
1.03 ± 0.01 s $\quad 104.89 \pm 0.01$ s

Example of percentage error: 97 cm $\pm 2\%$ $\qquad 12$ cm $\pm 20\%$
1.03 s $\pm 1\%$ $\quad 104.89$ s $\pm 0.01\%$

The rules for compounding measurements through addition or multiplication are based on the assumption that the worst possible coincidence of errors will occur. For instance, in adding 97 ± 2 cm to 12 ± 2 cm the sum could be as large as 113 or as small as 105 if the original values were off by the maximum amount *in the same direction*. Often there will be cancellation of errors in arriving at a compound quantity, because some of the original measurements will have values that are too high and others will have values that are too low. On p. 12 we show how this effect can be taken into account, but these second approximation rules are simpler and often satisfactory.

B. To find compound error in addition or subtraction, *add* the absolute errors.

$(97 \pm 2 \text{ cm}) + (12 \pm 2 \text{ cm}) = 109 \pm 4$ cm
$(104.89 \pm 0.01 \text{ s}) - (1.03 \pm 0.01 \text{ s}) = 103.86 \pm 0.02$ s
$(88 \pm 2 \text{ kg}) + (3.26 \pm 0.02 \text{ kg}) = 91 \pm 2$ kg

C. To find compound error in multiplication or division, *add* the *percentage* errors.

Area of square with sides 97 cm $\pm 2\% = 9400$ cm² $\pm 4\%$
$= 9400 \pm 400$ cm²

$(97 \text{ cm} \pm 2\%) \times (12 \text{ cm} \pm 20\%) = 1200$ cm² $\pm 20\%$
$= 1200 \pm 200$ cm²

$(104.89 \text{ s} \pm 0.01\%)/(1.03 \text{·s} \pm 1\%) = 100 \pm 1\% = 100 \pm 1$

The determination of the absolute error is a matter of judgment. How would one estimate that a length is 97 ± 2 cm? The error being assigned to this value has nothing to do with any mistake. There is no mathematical treatment that can automatically compensate for mistakes. The error limits of ± 2 cm may be assigned as the result of the observation of numerical data. Several people may have measured the length or one person may have done it several times, and all the values were found to be between 95 cm and 99 cm, with 97 as the average. Perhaps only one measurement was made with a meter stick marked every 10 cm, and the experimenter estimated that the length was about 97 cm and could not be more than 2 cm different. In this case, the instrument itself was the limiting factor. Perhaps the measurement was done with a tape that had markings down to millimeter size, but the experimenter did not trust the shrinking or streching of the tape closer than ± 2 cm. Perhaps the measuring instrument was precise and trustworthy, but the object being measured was moving and difficult to measure, or irregular so that a closer measurement was not justified. In all these cases, the assignment of reasonable error limits depends on the *judgment of the experimenter* based on the *instrument*, the *object*, and the *need for precision*. Note that it is *not* generally true that the absolute error is equal to some fraction of the smallest scale division of the measuring instrument. For instance, you could use a vernier caliper to measure the diameter of a piece of chalk, but because of the unevenness of the chalk the error would be closer to 2 mm than to 0.1 mm.

Note from the examples in A that the same absolute error in two measurements may yield radically different percentage errors. Note also that if only one significant figure is given in the absolute error, only one significant figure is justified in the percentage error. The third example in B illustrates the same sort of principle in the addition of errors. If there is only one significant figure in one of the errors, it seldom makes sense to end up with a compound error that contains two significant figures. As pointed out on p. 7, an exception to this rule is justified if the first digit of the final number is 1. Thus $(62 \pm 10) + (21 \pm 5) = 83 \pm 15$.

Although the numbers used as examples in this section displayed symmetrical errors, there is no necessity for the given value to lie halfway between the extremes. For example, suppose you were sure that a mass that was determined to be 102 grams (g) could not be less than 100 g but might be as large as 108 g. Its mass should then be given as $102\left(\begin{smallmatrix}+6\\-2\end{smallmatrix}\right)$ g.

Justification for the rule for compounding errors in multiplication can be obtained algebraically, graphically, and by the trial of extreme values. Suppose that an area with true length L and true width W is measured to have values $L + l$ and $W + w$, where l and w are the absolute error.

Algebraic: True area $= LW$

$$\text{Calculated area} = (L + l)(W + w) = LW + lW + wL + lw$$

$$\% \text{ error in area} = \frac{(\text{calculated area} - \text{true area})}{\text{true area}} 100$$

$$= \frac{lW + wL + lw}{LW} 100 = \frac{l}{L} 100 + \frac{w}{W} 100 + \frac{lw}{LW} 100$$

$$= (\% \text{ error in } L) + (\% \text{ error in } W)$$

$$+ (\text{comparatively small term})$$

1.4 ABSOLUTE AND PERCENTAGE ERRORS

If l/L and w/W are $\ll 1$, lw/LW is small compared with l/L or w/W.

Graphical:

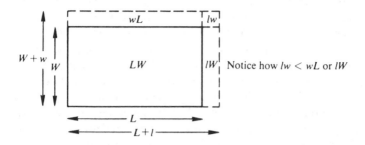

Notice how $lw < wL$ or lW

Numerical trial: If $L = 100 \pm 10$ cm and $W = 50 \pm 10$ cm, then $L = 100 \pm 10$ per cent and $W = 50 \pm 20$ per cent. The rule gives: Area = 5000 cm² $\pm$ 30 per cent or ± 1500 cm². If $L = 110$ cm and $W = 60$ cm, $A = 6600$ cm². If $L = 90$ cm and $W = 40$ cm, $A = 3600$ cm².

SAMPLES

1. What is the volume of a sphere that has a diameter of 6.2 ± 0.2 cm? (Volume $= \frac{4}{3}\pi r^3 = \frac{1}{6}\pi d^3 = 1.2 \times 10^2$ cm³ $\pm 10\%$.)

2. Carry out the algebraic demonstration for the other extreme case in which the measured values are $L - l$ and $W - w$. Does the rule still hold?

3. How long did an event last if it started at 3.4 ± 0.2 s and ended at 5.0 ± 0.2 s? What is the percentage accuracy?   (1.6 ± 0.4 s or $\pm 30\%$).

4. What is the volume of a cylinder with length 6.24 ± 0.01 cm and diameter 2.1 ± 0.1 cm?   (22 ± 2 cm³ or $\pm 10\%$)

5. What is the average velocity if a bullet travels 100.00 ± 0.1 m in 0.15 ± 0.01 s?   (670 ± 40 m/s or $\pm 7\%$)

1.5 DATA DISTRIBUTION CURVES—A THIRD APPROXIMATION TO ERROR ANALYSIS. The use of this third approximation to error analysis is justified only when certain experimental conditions and demands are met. If the formalism is applied blindly, as it often is, sophisticated precision may be claimed when it does not exist at all. The mathematical techniques are derived from and are concerned with statistical laws of probability. The actual measurements are assumed to cluster around an average value and to vary from that average in a way specified by the bell-shaped Gaussian distribution.

In Sections 5.1 and 4.6 we describe the Gaussian function and its properties. This particular distribution function occurs under very general conditions of chance determination. The major requirement is that if the probability of getting a particular value i is p_i and the probability of getting

REPORTING AND ANALYZING UNCERTAINTY

j is p_j, then the probability of getting the values i and j in sequence is the product $p_i p_j$. It is also necessary that the probability be maximum for one particular value. These conditions are frequently met by data subject to ordinary errors of measurement.

There are many reasons why the measurements may not fit the Gaussian function; there may be some cut-off or prejudice against low readings as opposed to high readings, producing a skewed distribution; there may be mistakes in readings or in instruments; the quantity being measured or the instruments may have periodic variations; the data sample may be so small that statistical fluctuations cause a warped distribution. Even when the *precision* (the scatter of data), determined by the statistical analysis of random errors, is correctly stated, the *accuracy* may be poor because of systematic errors or other types of mistakes.

If elaborate analysis of experimental error is necessary, it may be helpful to consult the following references:

Introduction to the Theory of Error, Yardley Beers (Addison-Wesley, Reading, Mass., 1958).

Experimentation and Measurement, W. J. Youden (Scholastic Book Service New York, 1962).

Experimental Measurements: Precision, Error and Truth, N. C. Barford (Addison-Wesley, Reading, Mass. 1967).

An Introduction to Error Analysis, John R. Taylor (University Science Books, Mill Valley, California, 1982).

If repeated measurements of the same quantity produce a distribution curve like the one shown in the diagram, it is misleading to use the maximum range as the error limits. The precision is really better than that. A reasonable criterion for the width of the uncertainty is σ, the standard deviation of the Gaussian function. Two thirds of the data points fall within the range between $\bar{x} + \sigma$ and $\bar{x} - \sigma$. The mean, $\bar{x}$, is the arithmetic average of the readings.

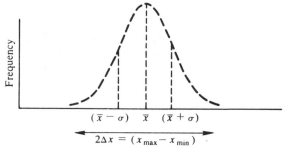

The standard deviation is sometimes called the *root-mean-square deviation*. It can be calculated from the data in a straightforward, though sometimes laborious, way. An example of such a calculation is given on p. 91.

$$\sigma = \sqrt{\frac{\sum_{i=1}^{n}(x_i - \bar{x})^2}{n}}$$

1.6 PROPAGATION OF ERRORS IN THE THIRD APPROXIMATION.

In the second approximation to error analysis we made the most pessimistic assumption about the way that errors in separate measurements might combine to yield a total error. If $x \pm \Delta x$ is to be added to $y \pm \Delta y$, it is possible that the true value could be as large as $x + y + \Delta x + \Delta y$ or as small as $x + y - \Delta x - \Delta y$. In some cases, however, it is possible that the errors will cancel to some extent so that the range of final error will not be $\pm(\Delta x + \Delta y)$. If there are sufficient data to justify the use of the standard deviation to express the error probabilities in the measurement of the separate variables, then the rules in this section can be used in calculating compound errors. One additional requirement is that the variables and their measurement be independent of each other. (Note Sample 3.)

A. The plausibility that there will be cancellation in combining errors can be seen by considering what happens if two data distribution curves are added to each other. The Gaussian peaks will overlap. A revealing way to think of the error combination is in terms of right-angle geometry, which the formula below suggests. This representation emphasizes the way in which a larger error dominates a smaller one.

If $s = x \pm y, \quad \sigma_s = \sqrt{\sigma_x^2 + \sigma_y^2}.$

1. $s = x + y \qquad x = 18.4 \pm 0.2 \quad y = 16.2 \pm 0.2$

$$\sigma_s = \sqrt{(4 \times 10^{-2}) + (4 \times 10^{-2})} = 0.3$$

$$s = 34.6 \pm 0.3$$

2. $s = x - y \qquad x = 42 \pm 2 \text{ cm}, \quad y = 20.12 \pm 0.02 \text{ cm}$

$$\sigma_s = \sqrt{(4) + (4 \times 10^{-4})} = 2$$

$$s = 22 \pm 2 \text{ cm}$$

B. The usefulness of repeated measurements of a quantity is illustrated in the example given below. Note that this relationship is valid only if each of the measurements contains sufficient data to justify the use of the standard deviation. (This relationship is actually just a special case of the rule given in A.)

$$\bar{\sigma} = \frac{\sigma}{\sqrt{n}}$$

Results of four measurements of p:

$p(1) = 83 \pm 2.5,$

$p(2) = 84 \pm 2.4,$

$p(3) = 81 \pm 2.7,$

$p(4) = 85 \pm 2.$

Then the average value is $p = 83 \pm 1.2.$

c. The rule for finding errors of products is just an extension of the rule used in the second approximation, but with the probability of cancellation of errors taken into account. The derivations of these particular rules plus descriptions of how to deal with more complicated situations (such as when the variables are not completely independent) are given in the references cited on p. 11.

If $p = xy$ or x/y, $\quad \sigma_p/p = \sqrt{(\sigma_x/x)^2 + (\sigma_y/y)^2}$.

1. $p = xy \qquad x = 21 \pm 0.5$ cm, $\quad y = 10 \pm 0.5$ cm,

$$\frac{\sigma_x}{x} = 2.5 \times 10^{-2}, \quad \frac{\sigma_y}{y} = 5 \times 10^{-2}$$

$$\frac{\sigma_p}{p} = \sqrt{(6.3 \times 10^{-4}) + (25 \times 10^{-4})} = 5.6 \times 10^{-2}$$

$$p = 210 \pm 12 \text{ cm}^2$$

2. $p = \dfrac{x}{y} \qquad x = 2.3 \pm 0.1$ radian (rad), $\quad y = 1.002 \pm 0.001$ s,

$$\frac{\sigma_x}{x} = 4 \times 10^{-2}, \quad \frac{\sigma_y}{y} = 1 \times 10^{-3}$$

$$\frac{\sigma_p}{p} = \sqrt{(16 \times 10^{-4}) + (1 \times 10^{-6})} = 4 \times 10^{-2}$$

$$p = 2.3 \pm 0.1 \text{ rad/sec}$$

SAMPLES

1.　By using the rule in A, show that the average standard deviation of n measurements, each with about the same standard deviation, is given by the rule in B. [Calculate the sum of the values and the sum of their standard deviations ($\sqrt{n}\ \sigma$). Then divide by n to get the mean and its standard deviation.]

2.　The number of photons per second coming through a filter is 1.06×10^4 $\pm 1 \times 10^2$. If a detector is gated with an on-time of 10.0 ± 0.1 milliseconds (ms), how many photons will it detect?　　(106 ± 1.4 photons/gate. How could there be an uncertainty of 1.4 photons when the number of photons must be an integer? What is the significance of the mathematical result?)

3.　The diameter of a sphere is measured to be 1.000 ± 0.002 cm. What is the volume?　　(Volume $= 0.524 \pm 0.003 \text{ cm}^3$.)

4.　The total reaction time of a relay is 38 ± 2 ms. The time for the contact arm to leave the base position is 20 ± 2 ms. How long is the part of the reaction when the arm is moving?　　(18 ± 3 ms.)

5.　The measured masses of five blocks are 390 ± 10 milligrams (mg), 460 ± 10 mg, 270 ± 10 mg, 540 ± 10 mg, and 420 ± 10 mg. What is the total mass?　　(2080 ± 22 mg).

1.7 ACCURACY, PRECISION, AND COMMON SENSE

A. Does an average of 10 readings, each with n significant figures, yield a value with $n + 1$ significant figures?

$$\bar{x} = \frac{\sum_i x_i}{n} = \frac{\begin{matrix} 18.4 + 19.2 + 18.5 + 18.9 + 18.8 + 18.5 \\ + 18.9 + 18.6 + 19.0 + 18.7 \end{matrix}}{10} \overset{?}{=} 18.75$$

It is commonly assumed that this is the case. The standard argument is that since the first figure after the decimal is significant for each reading, the first figure after the decimal should be significant for the sum (in this case, 187.5). Dividing by 10 does not change this significance, and therefore the average has an additional significant figure.

This situation illustrates the difficulty of laying down hard and fast rules about the analysis of data. In the example given, it may be that the experimenter could judge each reading within ± 0.05, but for one reason or another the average value did not always lie within that range. The spread of readings indicates that the figure after the decimal point was not significant in the original technical sense. Circumstances such as this are common in actual experiments. In general, there is no justification for citing an additional significant figure for an average value.

B. The error in a student experiment is *not* the difference between the student's experimental value and the textbook value. Error limits of individual measurements should be determined and cited by the experimenter. The compound error due to errors in the individual variables should then be calculated. If the error limits of the experimental value do not overlap the textbook value, there is evident reason to reassess the original judgments of possible error limits or to look for a mistake. An experimental value of 84 ± 20 is not in disagreement with a textbook value of 100. (Of course, if the assignment was to determine the value within 5 per cent, the experimenter has other problems.)

C. For a compound value made up of several individual values, do not seek more precision for any one value than is justified by the precision with which you know the others. If one value has a 10 per cent error, there is usually no point in obtaining another value to within 1 per cent. Notice especially how unimportant small errors become if they express standard deviations, linked with others through the square root of the sum of their squares. (In this case, the product of two values, one with a 10 per cent error and the other with a 3 per cent error, has an error of $\sqrt{10^2 + 3^2} = 10.4$ per cent, and not 13 per cent)

If the diameter of a cylinder is measured to within 2 per cent and the height to within 1 per cent, then the volume is known to within 5 per cent, assuming that the data were not sufficient to justify statistical procedures. The use of three significant figures for π (3.14) yields $\frac{1}{6}$ per cent error margins, which do not add to the overall error margins for the volume. The use of four significant figures would not be justified. There would also be no point in improving the precision of the measurement of the height, since the major contribution to the error is provided by the diameter measurement.

D. You have no doubt heard that ancient saw, "If a thing is worth doing, it is worth doing well." That, of course, is nonsense. If a thing is worth doing, it is worth doing *well enough for the purpose at hand.* To do it any better than that is surely silly and probably wrong. Do not think, however, that this realistic view makes life easier. The purpose at hand may require years of devoted and meticulous work. Furthermore, the individual is faced with the awful responsibility of using his head to determine the requirements of the problem. No rules exist.

Precision is expensive. In 1954 a particular cross-section value for a high-energy particle reaction was measured during the course of an afternoon to an accuracy of 15 or 20 per cent. During the following two years, three scientists and numerous technicians spent a considerable fraction of their time and over $100,000 to obtain that cross section to a ± 4.4 per cent error. There was good reason for obtaining that precision, and the probable cost and difficulty were carefully considered in advance. The very first response that anyone should make when faced with a task is: "For what purpose is this required?" Your procedure will often depend strongly on the answer.

What is the area of your front yard? Your choice of measuring instruments depends on the purpose for which the information is required. If you want to know the area so that you can buy lime in 80-lb bags at the store, then you can measure the yard by looking at it. Lime is cheap, you cannot buy a fractional bag, and the exact dosage is unimportant. If you want to know the area in order to buy grass seed at $1.00/lb, then you should pace out the yard and perhaps even use pencil and paper to check your arithmetic. Whether the yard is exactly rectangular, or whether your pace is 5 feet (ft) or 5 ft 3 in is unimportant. You certainly would not use a meter stick. (Where is the seed salesman who can advise you in terms of lb/m²?) If you want to measure the area of your yard for legal purposes, the assessor will probably insist on knowing the area to within 0.01 acre. A surveyor's transit is the appropriate measuring instrument.

1.8 PROPER CITATION OF ERROR LIMITS

See Appendix 11 on p. 256.

CHAPTER 2 UNITS AND DIMENSIONS

2.1 CHANGES OF UNITS. Both the technical world and our everyday world are bedeviled with a multitude of units used to measure the same quantity. Length is measured in meters, feet, centimeters, inches, kilometers, yards, miles, furlongs, spans, fathoms, microns, Ångstroms, Fermi's, millimeters, rods, chains, light-years, parsecs, barleycorns, and probably many others. Some of these units survive as accidents of history, but many are used because they are convenient for some particular trade. It is useful to have a standard unit about the same size as the object being measured.

It is often necessary to change units for consistency in calculations. For simple cases this can be done without formal bookkeeping, but there is a foolproof way of keeping track of the changes. The quantity to be converted should be written in the original numbers with all units specified in words or symbols, e.g., 60 miles/hr. The units are then treated like algebraic quantities that can be multiplied, divided, and canceled. The guiding rule for conversion of the units is to multiply the original value by fractions that are equal to 1. For example, let's change 60 miles/hr to feet/second. We need the following identities: 5280 ft = 1 mile; 60 minutes (min) = 1 hr; 60 s = 1 min.

$$\frac{60 \text{ miles}}{\text{hr}} \times \frac{5280 \text{ ft}}{1 \text{ mile}} \times \frac{1 \text{ hr}}{60 \text{ min}} \times \frac{1 \text{ min}}{60 \text{ s}} = \frac{60 \times 5280}{60 \times 60} \frac{\text{ft}}{\text{s}} = 88 \text{ ft/s}$$

This conversion factor is frequently useful both in the technical and the everyday world. Notice that each of the multiplying fractions has the value 1. The fraction 5280 ft/1 mile was used, rather than 1 mile/5280 ft because we wanted to cancel the miles in the original numerator. The only units left were the ones desired: feet in the numerator and seconds in the denominator. Spelling out the units in this fashion keeps an automatic check on whether a conversion factor (such as 5280) should be multiplied or divided.

For another example, let us convert furlongs per fortnight to speed in cm/s.

$$\frac{1 \text{ furlong}}{\text{fortnight}} \times \frac{220 \text{ yd}}{1 \text{ furlong}} \times \frac{3 \text{ ft}}{1 \text{ yd}} \times \frac{12 \text{ in.}}{1 \text{ ft}} \times \frac{2.54 \text{ cm}}{1 \text{ in.}} \times \frac{1 \text{ fortnight}}{14 \text{ days}}$$

$$\times \frac{1 \text{ day}}{24 \text{ hr}} \times \frac{1 \text{ hr}}{3600 \text{ s}} = 1.66 \times 10^{-2} \text{ cm/s}$$

Evidently 1 furlong/fortnight is a snail's pace. This conversion factor is little used.

SAMPLES

1.   Change atmospheric pressure from 14.7 lb/in.^{2} to dynes/cm^2. One pound $= 4.45 \times 10^5$ dynes. One meter $= 39.37$ in. $(1.01 \times 10^6 \text{ dynes/cm}^2.)$

2.   The gas constant in the equation $PV = nRT$ is $R = 8.31$ joules(J)/mole-°K. Find the value of R in (eV)/molecule-°K. $1.6 \times 10^{-19} \text{ J} = 1 \text{ eV}$ (electron volts); 6×10^{23} molecules $= 1$ mole. $[(8.65 \times 10^{-5} \text{ eV})/\text{molecule-°K.}]$

3.   Given that the density of rock is 2.5 g/cm^3, find the density of rock in kg/m^3. $(2.5 \times 10^3 \text{ kg/m}^3)$.

4.   Change 1 light-year to meters. (A light-year is the distance light travels in 1 year.) The speed of light is 3.0×10^8 m/s. $(9.5 \times 10^{15} \text{ m.})$

5.   A physics student using a tickertape and a doorbell clapper that can oscillate 14 times per second found the freefall acceleration of a lead weight to be 5.0 cm/dot^2. What was g equal to in meters per square second?   (9.8 m/s^2.)

6.   An electric fan has been "stopped" by a circular strobe rotating at 4 revolutions per second. The strobe disc has 12 slits in it and the fan has one blade painted red. What is the speed of the fan in revolutions per minute (rpm) if this is the slowest speed before double images begin to appear? (2880 rpm.)

2.2 DIMENSIONS.   Under no circumstances can we justify an equation that says

4 horses = 5 cows + 5 sheep

It is conceivable, however, that the *cost* of these animals might be equated.

4 horses × $500/horse = 5 cows × $300/cow + 5 sheep × $100/sheep

Each term in this second equation is a certain number of dollars. In every equation each term must describe the same quantity as every other term.

Frequently, the terms may not look alike. For instance, the final velocity of an object starting out with velocity zero and subject to constant accelaration, a, for a distance, s, is

$$v_f^2 = 2as$$

The left side is a velocity squared; the right side is the product of an acceleration and a distance. Each of these, however, is composed of more primitive dimensions. The basic three dimensions are length (L), mass (M), and time (T). Velocity consists of a length divided by a time: L/T. Acceleration has the dimension L/T^2. The dimensions of the terms in the equation above are thus

$$\left(\frac{L}{T}\right)^2 = \frac{L}{T^2}L$$

Dimensionally, the terms are the same.

This simple requirement provides a quick check on the consistency of results in a complicated derivation, and also is the basis of powerful methods of analysis, particularly in complex situations requiring gross parameters, such as in hydrodynamics. As a trivial example of the method of dimensional analysis, what is the final velocity under constant acceleration if there is an initial velocity v_0? The extra term must also have the dimensions of $(L/T)^2$. The simplest possibility is to introduce v_0^2:

$$v_f^2 = v_0^2 + 2as$$

This makes sense because of an additional argument. If $a = 0$, then v_f^2 must $= v_0^2$.

Here are two examples of dimensional analysis used to derive the functional dependence of one variable on others.

1. Under constant acceleration, the distance traveled from rest is a function of a and t. What is the function?

   $$s = f(t, a)$$
   $$s = Kt^n a^r$$

   Dimensionally, $L^1 = T^n(L/T^2)^r$. This can be satisfied if $r = 1$ and $n = 2$. Therefore, $s = Kat^2$.
   Dimensional analysis cannot give the value of the constant.

2. Is the period of a simple pendulum

   $$T = 2\pi\sqrt{\frac{l}{g}} \quad \text{or} \quad 2\pi\sqrt{\frac{g}{l}}?$$
   $$T = f(l, g) = Kl^n g^r$$

   Dimensionally, $T^1 = L^n(L/T^2)^r$. This can be satisfied if $n = -r$ and $r = -\frac{1}{2}$. Therefore, $T = Kl^{1/2}g^{-1/2} = K\sqrt{l/g}$. Once again, the method is incapable of giving the constant.

Dimensional analysis can be used to discover much more complicated functional relationships. For an elementary but very complete treatise on the subject see *Dimensional Analysis* by H. E. Huntley, Dover Publications, New York, 1967.

Routine use of the method simply as a check on derivations can pay off in time saved. For instance, note that the arguments of all sines, exponents, etc., must be dimensionless. (Angles are dimensionless, since $\theta = $ arc/radius $= L/L$.) For a traveling wave,

$$y = A \sin 2\pi\left(\frac{t}{T} - \frac{x}{\lambda}\right) = A \sin \omega\left(t - \frac{x}{v}\right)$$

In the first expression each term is dimensionless: T/T, L/L. In the second expression each term in the parentheses must have the dimensions of T, since ω is a frequency with the dimension of T^{-1}. The dimension of x/v is $L/LT^{-1} = T$.

For a more complicated case, consider the exponents:

$$E = E_0 e^{-\sqrt{\sigma\omega\, 2\epsilon}\,(x/c)} e^{i\omega[t - \sqrt{\sigma\, 2\epsilon\omega}\,(x/c)]}$$

Are those complicated expressions dimensionless? (For dimensions of electrical quantities see the table on p. 20.)

  σ is conductivity, $L^{-2}T$
  ω is angular frequency, T^{-1}
  ϵ is permittivity of free space, $L^{-2}T^2$

Dimensions of

$$\sqrt{\frac{\sigma\omega}{2\epsilon}}\frac{x}{c} \longrightarrow \sqrt{\frac{L^{-2}TT^{-1}}{L^{-2}T^2}}\frac{L}{LT^{-1}} \longrightarrow \sqrt{T^{-2}}\,T \longrightarrow 1$$

The second exponential can be checked directly from the results of the first.

The dimensions of most quantities are just multiples of L, M, and T. For work in thermodynamics it is convenient to consider temperature a separate dimension, denoted by θ. Electromagnetic quantities have more complicated dimensions that include fractional values unless charge, Q, is taken as a fundamental dimension. The system has considerable arbitrariness connected with it, since other dimensions could have been chosen as fundamental. It might be argued, for instance, that velocity is a fundamental dimension, since the velocity of light is a universal constant.

The following tables list the dimensions of some common quantities:

Angle, 0	Frequency, T^{-1}
Area, L^2	Viscosity, $ML^{-1}T^{-1}$
Density, ML^{-3}	Acceleration, LT^{-2}
Moment of inertia, ML^2	Force, MLT^{-2}
Momentum, MLT^{-1}	Pressure, $ML^{-1}T^{-2}$
Angular momentum, ML^2T^{-1}	Surface tension, MT^{-2}

Torque, L^2MT^{-2}
Energy, ML^2T^{-2}
Power, ML^2T^{-3}
Temperature, Θ
Specific heat, 0

Heat (energy), ML^2T^{-2}
Heat conductivity, $MLT^{-3}\Theta^{-1}$
Heat capacity per mass, $L^2T^2\Theta^{-1}$
Entropy, $ML^2T^{-2}\Theta^{-1}$

With Q as a fundamental dimension		In Terms of M, L, T only
Q	Electric charge (q)	$L^{1/2}M^{1/2}$
$T^{-1}Q$	Electric current (i)	$L^{1/2}M^{1/2}T^{-1}$
$ML^2T^{-2}Q^{-1}$	Potential difference (V)	$M^{1/2}L^{3/2}T^{-2}$
$ML^2T^{-1}Q^{-2}$	Resistance (R)	LT^{-1}
$M^{-1}L^{-2}T^2Q^2$	Capacitance (C)	$L^{-1}T^2$
ML^2Q^{-2}	Inductance (L)	L
MLQ^{-2}	Permeability (μ)	0
$M^{-1}L^{-3}T^2Q^2$	Permittivity (ϵ)	$L^{-2}T^2$
$L^2T^{-1}Q$	Magnetic dipole moment (μ)	$L^{5/2}M^{1/2}T^{-1}$
$MT^{-1}Q^{-1}$	Magnetic induction (B)	$M^{1/2}L^{-1/2}T^{-1}$
$MLT^{-2}Q^{-1}$	Electric field (E)	$M^{1/2}L^{1/2}T^{-2}$
$M^{-1}L^{-3}TQ^2$	Conductivity (σ)	$L^{-2}T$
$L^{-2}T^{-1}Q$	Current density (j)	$M^{1/2}L^{-3/2}T^{-1}$

SAMPLES

1. Check the dimensionality of the following well-known formulas:

a. $F_{\text{grav}} = G(m_1 m_2/r^2)$ (What must be the dimensions of G?)

b. $F_{\text{centripetal}} = mv^2/r = m\omega^2 r$

c. $F_x = -dU/dx$ (where U is potential energy)

d. torque $= dL/dt$ (where L is angular momentum.)

e. $dU = T\,dS - P\,dV$ (where U is internal energy, T is temperature, S is entropy, P is pressure, and V is volume)

f. $\omega = mgR/L$ (where ω is precession frequency)

g. $r = n^2\epsilon h^2/\pi me^2$ (where ϵ is permittivity; h is Planck's constant, which is an angular momentum; n is an integer; r, m, and e are radius of orbit, mass, and charge of an electron in a hydrogen atom)

h. $L(d^2q/dt^2) + R(dg/dt) + (1/C)q = 0$ (where L is inductance, R is resistance, and C is capacitance)

i. $\omega = \sqrt{1/LC - (R/2L)^2}$

j. $F = \mu li_1 i_2/2\pi r$ (Ampere's law, where l is length of wires carrying i_1 and i_2 amperes, separated by distance, r)

CHAPTER3GRAPHS

There are some general conventions about drawing graphs, most of them sensible. These are illustrated in the following pages. Adherence to the conventions is easy and is guaranteed to save time.

Data should usually be graphed as soon as they are obtained. Gross mistakes are not always apparent when the data are entered in tabular form. If, however, a graph of the data is being plotted while it is coming in, a point that is radically out of line is immediately obvious.

A. A graph should have a title that explains the functional representation and the coordinates should be labeled with proper units.

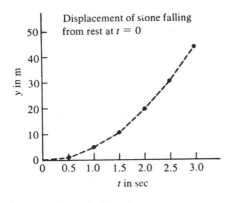

y is a function of t. Since t is the independent variable, it is plotted along the horizontal axis, the abscissa. The dependent variable, y, is plotted along the vertical axis, the ordinate.

B. Choose the coordinate scale so that the functional behavior can be clearly shown. In each of these first two examples, most of the graph is left unused. Compare them with the graph of the same data in A.

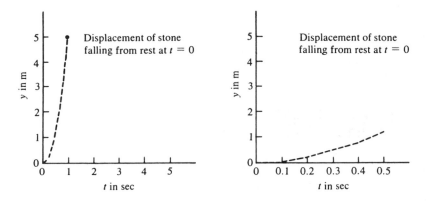

Note the different appearance of graphs of the same function. By skillful choice of scale you can mask or enhance the appearance of changes in a variable.

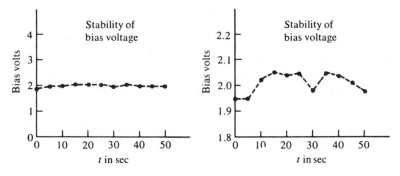

C. Connecting the data points on a graph with a curve is justified only if a functional dependence is assumed and if there is reason not to expect fine

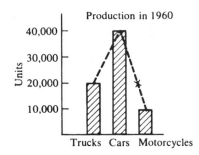

The dotted line is meaningless. (Would the point between cars and motorcycles represent tricycles?)

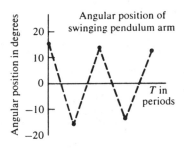

Joining these data points with straight lines clearly leads to false conclusions.

structure between points. Bar graphs or histograms frequently do not meet these conditions.

D. The slope of a curve at a point is equal to the slope of the tangent line at that point. The tangent line on a graph can be determined with fair precision by using a plastic ruler and a pencil. Place the pencil point on the graph point where the slope is to be found. Rotate the edge of the ruler around that point until the gaps between straight edge and curve on either side of the point appear to be equal. Draw the straight line. Choose any convenient length of straight line for the slope.

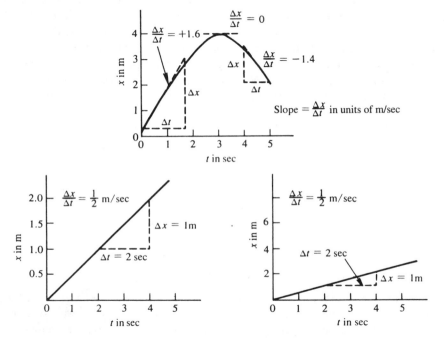

Note that the slope of both of these lines is the same. The different appearance is due to the different ordinate scales.

E. Graph an equation of a straight line:

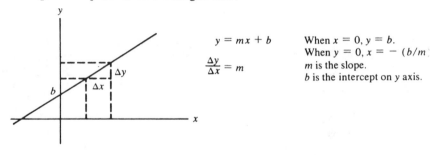

$$y = mx + b$$

$$\frac{\Delta y}{\Delta x} = m$$

When $x = 0, y = b$.
When $y = 0, x = -(b/m)$
m is the slope.
b is the intercept on y axis.

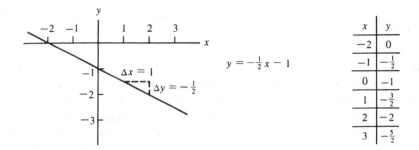

x	y
-2	0
-1	$-\frac{1}{2}$
0	-1
1	$-\frac{3}{2}$
2	-2
3	$-\frac{5}{2}$

$$y = -\tfrac{1}{2}x - 1$$

F. Error representation on graphs: Each data "point" should be represented as a region of possible values defined by the error limits within which the value may exist. This graphical representation of the data and possible errors provides an immediate check on the consistency of the data and indicates if any points should be measured again. The range of possible slopes is evident, and it is also easy to see how precision of the slope value might be improved by measuring only certain points with increased precision.

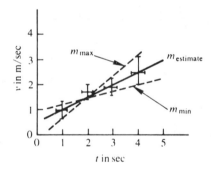

In this particular example, velocity was measured each second as well as could be gauged, which was ± 0.2 s. The velocity measurements were accurate to ± 20 per cent, making the absolute errors greater for the larger velocities.

G. Producing straight-line graphs by forming linear functions:

$$y = \tfrac{1}{2}g_0 t^2$$

$y \propto t^2$ yielding parabola when plotted vs. t

Displacement of stone falling from rest at $t = 0$

y	t	t^2
0	0	0
1.25	0.5	0.25
5.0	1.0	1.0
20.0	2.0	4.0
31.5	2.5	6.25
45	3.0	9.0

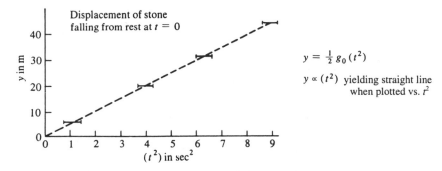

Displacement of stone falling from rest at $t = 0$

$y = \frac{1}{2} g_0 (t^2)$

$y \propto (t^2)$ yielding straight line when plotted vs. t^2

A linear or straight-line function can be identified very accurately on a graph. No one can judge visually whether the first graph is a plot of a parabola or some higher power curve. Furthermore, it is hard to notice if any particular data point is inconsistent with the assumed functional dependence. A straight line, however, can be judged by eye with great precision. (The precision is enhanced by sighting along the line with the eye as close as possible to the paper and to the projection of the line.) In the case above, y is proportional to t^2 and so a plot of y versus t^2 yields a straight line. Note that the error limits for t^2 will be larger than for t. If time can be measured to ± 0.1 s there will be ± 10 per cent error in t at $t = 1.0$ s, and ± 20 per cent error in t^2. At $t = 3.0$ s there will be ± 3 per cent error in t and ± 6 per cent error in t^2. Thus the error bar for the data region at $(t^2 = 9)$ extends from $(t^2 = 8.5)$ to $(t^2 = 9.5)$.

Functional dependence of experimental data can sometimes be determined by this graphical test. Draw up a chart of values for the possible powers of dependence:

x	y	y^2	y^3

.

Plot x versus y, x versus y^2, x versus y^3, etc., and judge which produces the best straight line. Note that the graph is plotted and the coordinates are constructed in terms of the values of the manufactured variable. That is, on the second graph above, the point at $t = 2$ s becomes the value at $t^2 = 4$.

H. Functional dependence that produces a straight line on semilog graph:

$y = ac^x$

$\log_{10} y = \log_{10} a + x \log_{10} c$

Take the log of both sides. Since $\log_{10} a$ and $\log_{10} c$ are constants, the equation can be written

$\log_{10} y = mx + b$ where $m = \log_{10} c$ and $b = \log_{10} a$

If $\log_{10} y$ is plotted versus x, it will yield a straight line with slope m. For instance, plot $\log_{10} y = 0.1x + 1$:

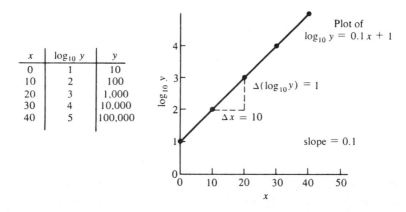

x	$\log_{10} y$	y
0	1	10
10	2	100
20	3	1,000
30	4	10,000
40	5	100,000

Plot of
$\log_{10} y = 0.1x + 1$

$\Delta(\log_{10} y) = 1$

$\Delta x = 10$

slope $= 0.1$

To eliminate the trouble of looking up the logs, use semilog paper.

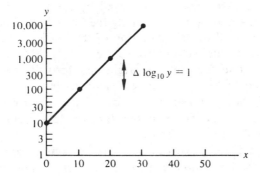

$\Delta \log_{10} y = 1$

The slope of this line has no meaning until Δy is translated to $\Delta \log_{10} y$. An easy way to make the translation is to note that each vertical centimeter on the graph paper represents $\Delta \log_{10} y = 1$.

An example of the use of such a graph is the plot of radioactive decay: $n = N_0 e^{-\lambda t}$. The number of particles left at time t is n; the original number of particles at $t = 0$ is N_0; $0.69/\lambda$ is the half-life, τ, the time during which half of any original number of particles decays. The number of particles decaying per second is $\Delta n/\Delta t = \lambda N_0 e^{-\lambda t} = \lambda n$. Since the decay rate is proportional to the number of particles left, a graph of either one versus time is equivalent to a graph of the other.

$$\log_{10} n = \log_{10} N_0 - \lambda t \log_{10} e = \log_{10} N_0 - 0.434 \lambda t$$

t (sec)	n (particles left)
0	100,000
100	24,700
200	6,100
300	1,500
400	330

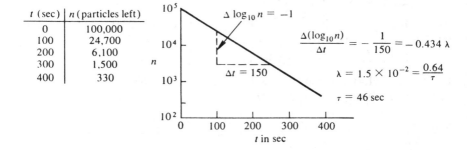

$\Delta \log_{10} n = -1$

$\Delta t = 150$

$\dfrac{\Delta(\log_{10} n)}{\Delta t} = -\dfrac{1}{150} = -0.434 \lambda$

$\lambda = 1.5 \times 10^{-2} = \dfrac{0.64}{\tau}$

$\tau = 46$ sec

I. Functional dependence that produces a straight line on a log–log graph:

$$y = ax^m$$

Take the log of both sides.

$$\log_{10} y = \log_{10} a + m \log_{10} x$$

Since $\log_{10} a$ is a constant, the equation has the form

$$\log_{10} y = m(\log_{10} x) + b$$

If $\log_{10} y$ is plotted versus $\log_{10} x$, it will yield a straight line with slope m. To eliminate the trouble of looking up the logs, use log–log paper:

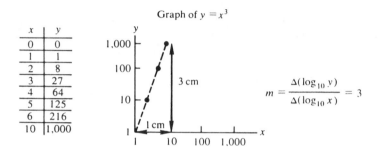

Graph of $y = x^3$

x	y
0	0
1	1
2	8
3	27
4	64
5	125
6	216
10	1,000

$$m = \frac{\Delta(\log_{10} y)}{\Delta(\log_{10} x)} = 3$$

Since the scales of $\log_{10} y$ and $\log_{10} x$ are equal, the actual geometrical slope of the line on the graph paper is equal to the slope of the equation.

Sometimes if one variable is a power function of another variable, the actual power dependence can be found faster by plotting on log–log paper than by plotting y versus x^2, y versus $x^{2.5}$, y versus x^3, y versus $x^{3.5}$, etc., to determine the best straight-line fit.

J. Fitting the best straight line

As shown in comment F, there is a range of straight lines that can be drawn so that they pass within the error regions of all the data. Under most experimental circumstances, the line of best fit and the range of possible lines can be judged visually by moving a straight edge over the points, assuming that there is a linear functional dependence. There is a formal procedure for obtaining the best-fit equation ($y = mx + b$) that minimizes the y deviations from the line. Its use is justified only when each datum point has been calculated from the raw data with statistical procedures. In other words, the formal line fitting should be used only if the third approximation to error analysis is required.

In this procedure, called the *method of least squares*, it is assumed that the major errors are in y, the dependent variable. A particular point, y_i, differs from the line at that point by $\Delta y_i = y_i - y = y_i - (mx_i + b)$.

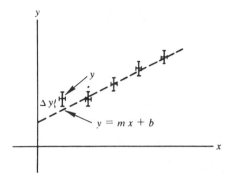

The straight-line parameters, m and b, should be chosen to minimize $\sum_{i=1}^{k} (\Delta y_i)^2$. A simple derivation (given, among other places in *Introduction to the Theory of Errors* by Y. Beers, p. 40) leads to these values:

$$m = \frac{k \sum (x_i y_i) - \sum x_i \sum y_i}{k \sum x_i^2 - (\sum x_i)^2} \tag{3-1}$$

$$b = \frac{\sum x_i^2 \sum y_i - \sum x_i \sum x_i y_i}{k \sum x_i^2 - (\sum x_i)^2} \tag{3-2}$$

The number of data points is k. Note that the denominator of m is the same as that of b. There is a real difference between $\sum x_i^2$ and $(\sum x_i)^2$ and similarly between $\sum (x_i y_i)$ and $\sum x_i \sum y_i$. For instance, if in the first case $x_1 = 1$ and $x_2 = 2$, then $\sum x_i^2 = (1)^2 + (2)^2 = 5$; $(\sum x_i)^2 = (1 + 2)^2 = 9$.

SAMPLES

1. Graph the following data showing the position of an object as a function of the time. Note that the graph makes information apparent that is not obvious in the columnar data.

$t(s)$	$x(m)$
0 ± 0.2	0 ± 0.1
1 ± 0.2	2 ± 0.1
2 ± 0.2	8 ± 0.2
3 ± 0.2	14 ± 0.2
4 ± 0.2	32 ± 0.3
5 ± 0.2	50 ± 0.4

2. Graph the following data showing the position of an object as a function of the time. Find graphically the velocity of the object at $t = 1$, $t = 5$, and $t = 8.5$ s.

t(s)	x(m)
0 ± 0.1	2.00 ± 0.03
2 ± 0.1	1.90 ± 0.03
4 ± 0.1	1.65 ± 0.03
6 ± 0.1	1.18 ± 0.03
8 ± 0.1	0.60 ± 0.03
9 ± 0.1	0.33 ± 0.03
10 ± 0.1	0.01 ± 0.03

3.   An object is moving with constant velocity at 4 m/s and at $t = 0$ was at $x = -15$ m. Graph its motion.

4.   Graph the following data and then find the algebraic relationship between x and y by plotting y versus various powers of x. The relationship is simple.

x	y
0 ± 0.1	0 ± 0.2
1 ± 0.1	1.9 ± 0.2
2 ± 0.1	2.7 ± 0.2
3 ± 0.1	3.2 ± 0.2
4 ± 0.1	4.0 ± 0.2
5 ± 0.1	4.4 ± 0.2
6 ± 0.1	5.1 ± 0.2

5.   Graph the following data on a semilog paper and find the decay half-life.

t(s)	n(no./s)	t(s)	n(no./s)
0	9800 ± 100	60	635 ± 25
10	6140 ± 79	70	385 ± 20
20	3840 ± 63	80	245 ± 16
30	2455 ± 50	90	170 ± 13
40	1540 ± 39	100	90 ± 10
50	1015 ± 32		

6.   Find the algebraic relationship between x and y by plotting the following data on log–log graph paper.

x	y
0	0
1	1.50
2	4.25
3	7.81
4	12.00
5	16.80
6	22.10

CHAPTER4 THE SIMPLE FUNCTIONS OF
APPLIED MATH

Many physical phenomena can be described in terms of a very few simple math functions. Even when the description is not exact, the simple functions are often good approximations over a limited range. It is particularly gratifying to believe that nature is so constructed that these few functions describe essentially all the phenomena of interest in introductory science courses. Alternatively, one would have to believe that introductory science deals only with those topics that can be handled with simple math.

In this section we shall describe these simple functions—naming them, defining them algebraically, graphing them, illustrating their use, and demonstrating special properties, such as their slopes.

4.1 POWER FUNCTIONS: $y = kx^n$.

A. Linear: $y = kx^1$ or, in general, $y = mx + b$ (4–1)

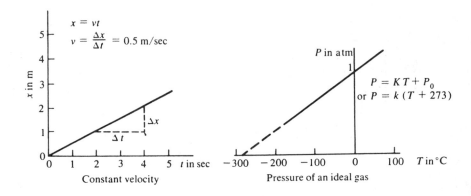

$x = vt$

$v = \dfrac{\Delta x}{\Delta t} = 0.5$ m/sec

Constant velocity

$P = KT + P_0$
or $P = k(T + 273)$

Pressure of an ideal gas

Note that it is not true in general that "if you double x you double y." This proportionality property is true only of the increments in each variable: $\Delta y = m\Delta x$.

For small increments, linearity is often a good approximation for many functions. You can judge qualitatively how good the approximation is by sliding a straight edge along the graph of the function.

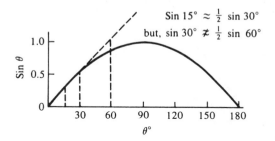

B. Quadratic: $y = kx^2$ or, in general, $y = ax^2 + bx + c$ (4-2)

(Note that this is also the equation for a parabola.)

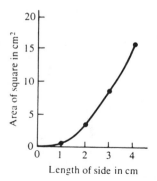

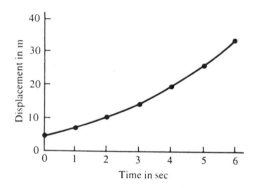

The slope of the quadratic is linear: $dy/dx = 2ax + b$. Note the relationships among displacement, velocity, and acceleration of a constantly accelerating object.

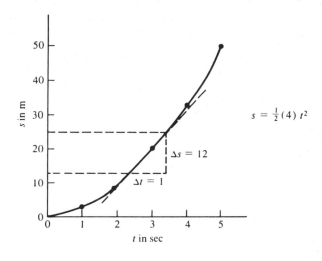

$$s = \tfrac{1}{2}(4)\, t^2$$

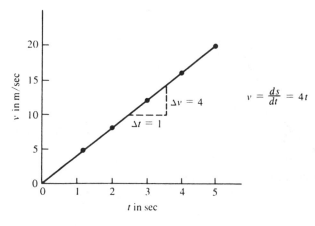

$$v = \frac{ds}{dt} = 4t$$

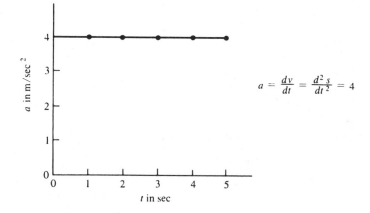

$$a = \frac{dv}{dt} = \frac{d^2 s}{dt^2} = 4$$

Relationship between quadratic and potential energy wells.   Aside from the dependence of areas on the square of linear dimensions, and the relationship between displacement and time with constant acceleration, the quadratic also represents the potential energy well in which many bound objects oscillate. Even when the potential function is not parabolic (quadratic), it can usually be approximated by the quadratic for small displacements from equilibrium. The slope of the potential energy function is related to the restoring force exerted by the system on the bound object:

$$F_{\text{restoring}} = -\frac{dU}{dx} \qquad (4\text{-}3)$$

If the potential energy has a quadratic dependence on the displacement,

$$U = -U_0 + \tfrac{1}{2} kx^2 \quad \text{and} \quad F_{\text{restoring}} = -kx \qquad (4\text{-}4)$$

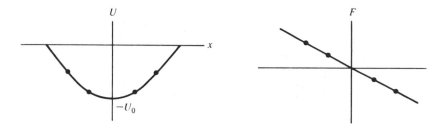

The restoring force is linear. It is zero at the equilibrium point, where $x = 0$ and the potential energy is a minimum. For a positive displacement, the restoring force is negative, forcing the object back to smaller values of x. For a negative displacement, the restoring force is positive. An essential feature is that the restoring force is proportional to the displacement, giving rise to simple harmonic motion (described in the section on the sinusoidal functions).

As described in the section on series, most functions can be approximated by

$$U(x) = a_0 + a_1 x + a_2 x^2 + a_3 x^3 + \cdots$$

For *small* displacements, x, around the equilibrium point, most potential wells can be approximately described by the first two even terms: $U(x) \approx a_0 + a_2 x^2$. (An odd term, such as the linear term $a_1 x$, would imply an asymmetric potential.) Consequently, small oscillations are subject to the mathematical model of the quadratic, yielding a linear restoring force and simple harmonic oscillations.

33

C. Square root: $y = kx^{1/2} = k\sqrt{x}$

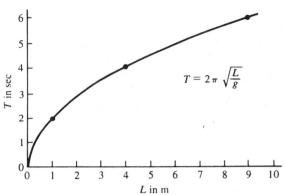

$$T = 2\pi \sqrt{\frac{L}{g}}$$

Period of simple pendulum vs. length

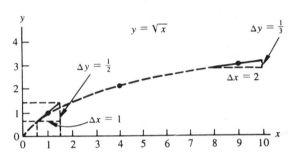

$y = \sqrt{x}$ $\Delta y = \frac{1}{3}$

$\Delta y = \frac{1}{2}$ $\Delta x = 2$

$\Delta x = 1$

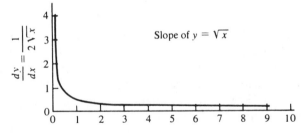

Slope of $y = \sqrt{x}$

$\frac{dy}{dx} = \frac{1}{2\sqrt{x}}$

D. Cubic and higher powers: $y = kx^3$

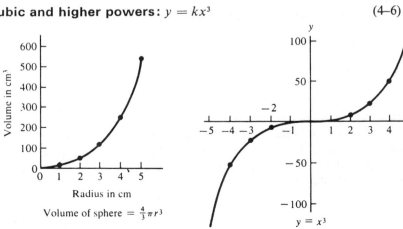

Volume in cm³

Radius in cm

Volume of sphere $= \frac{4}{3}\pi r^3$

$y = x^3$

E. Comparison of curves for $y = x^n$, with positive n

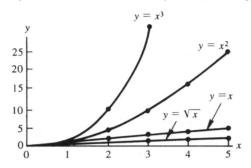

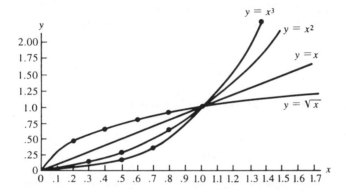

F. Inverse: $y = kx^{-1} = k/x$ (4–7)

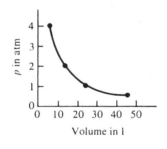

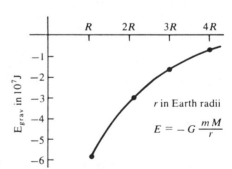

Pressure vs. volume for one
mole of ideal gas at $0°C$

Potential energy of one kg
in gravitational field of earth

Any influence (such as light intensity or electric field) that is emitted
uniformly from a *cylindrical* or *line* source (and does not decay and is not
absorbed) falls off in intensity according to the inverse first power. Here is
the geometrical reason for this.

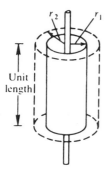

All the energy flow through the cylindrical shell (see the diagram) at r_1 must also go through the shell at r_2. The surface area of the curved part of a shell per unit length is

$$A = 2\pi r.$$

The energy flow *per area* (the intensity) is equal to

$$\frac{E}{A} = \frac{E}{2\pi r},$$

where E is the energy flow per unit length from the source. The assumption of uniform radial flow is good only when source length $\gg r$.

The slope of $y = kx^{-1}$ is

$$\frac{dy}{dx} = -kx^{-2}.$$

Since the restoring force owing to potential energy is

$$F_x = \frac{dE}{dx},$$

the gravitational force at any radius equals the negative value of the slope of the potential energy curve at that point. Hence

$$F_{\text{grav}} = -\frac{dE_{\text{grav}}}{dr} = -G\frac{mM}{r^2} \qquad (4\text{--}8)$$

The negative sign indicates that the radial force is toward the center. An inverse *first* power potential energy function yields an inverse *square* force function.

G. Inverse square: $y = kx^{-2} = k/x^2$ (4-9)

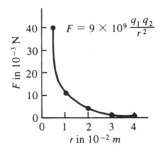

$$F = 9 \times 10^9 \frac{q_1 q_2}{r^2}$$

Repulsive force between two tiny positively charged spheres with 10^{-8} Coulombs each

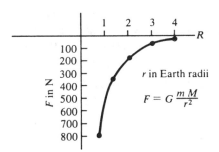

r in Earth radii

$$F = G \frac{m\,M}{r^2}$$

Attractive force between 80 kg man and 6×10^{24} kg planet

Any influence (such as light intensity or electric field) that is emitted uniformly from a *spherical* or *point* source (and does not decay and is not absorbed) falls off in intensity according to the *inverse square* of the distance to the center. Here is the geometrical reason for this. All the energy flow through the spherical shell at r_1 (see the diagram) must also go through the shell at r_2. The surface area of a spherical shell is $A = 4\pi r^2$. The energy flow *per area* (the intensity) is equal to $E/A = E/4\pi r^2$, where E is the energy flow from the source. Of course, the inverse-square rule applies only when the flow of energy (or spread of influence) is radial and isotropic, and in the region beyond the radius of the source.

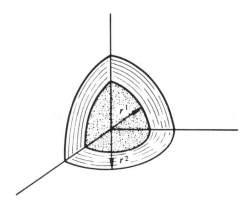

The slope of $y = kx^{-2}$ is $dy/dx = 2kx^{-3}$. The inverse cube represents the dependence of force on the distance of a dipole source. This is because the field at a distance from a dipole is the difference between the positive and negative inverse-square fields of the two poles. Along the line of the dipole,

4.1 POWER FUNCTIONS: $y = kx^n$

$$F = \frac{k}{(r - \frac{1}{2}\Delta r)^2} - \frac{k}{(r + \frac{1}{2}\Delta r)^2} \longrightarrow \frac{2k \, \Delta r}{r^3} \tag{4-10}$$

Since the tidal effects are caused by the *differential* pull of the moon or sun on surface water and the center of earth, the tides depend on the inverse cube of the distance between the earth and the moon or sun.

H. General equation for slope of power function:

$$y = kx^n \tag{4-11}$$

$$\text{slope} = \frac{dy}{dx} = nkx^{n-1}$$

$y = mx + b$ $\qquad\qquad \frac{dy}{dx} = m$

$y = a_0 + a_1x + a_2x^2$ $\qquad\qquad \frac{dy}{dx} = a_1 + 2a_2x$

$y = a_0 + a_1x + a_2x^2 + a_3x^3$ $\qquad \frac{dy}{dx} = a_1 + 2a_2x + 3a_3x^2$

$y = k\sqrt{x} = kx^{1/2}$ $\qquad\qquad \frac{dy}{dx} = \frac{1}{2}kx^{-1/2} = \frac{k}{2\sqrt{x}}$

$y = kx^{-1/2}$ $\qquad\qquad \frac{dy}{dx} = -\frac{1}{2}kx^{-3/2}$

$y = kx^{-1}$ $\qquad\qquad \frac{dy}{dx} = -kx^{-2}$

$y = kx^{-2}$ $\qquad\qquad \frac{dy}{dx} = -2kx^{-3}$

SAMPLES

1. Draw a precise graph of $y = x^2$. Find the slopes at various points, using a ruler to draw tangent lines. Plot these values to demonstrate that the graph of the slope of the quadratic is linear.

2. Draw a precise graph of $y = 1/x$. Find the slopes at various points, using a ruler to draw tangent lines. Plot these values to demonstrate that the graph of the slope of the inverse first power is an inverse-square function.

3. Plot the following data and determine the approximate power function dependence by examining the graph and trying to fit plausible functions.

x	y
0	0
1	1.00
2	1.68
3	2.28
4	2.83
5	3.35
6	3.83

THE SIMPLE FUNCTIONS OF APPLIED MATH

4. Over what range in miles at the earth's surface is it a good approximation (within 5 per cent) that $\Delta E_{grav} = mgh$, where m is the mass of an object, g is the planetary gravitational constant, and h is the height above the surface defined as zero potential? ΔE_{grav} is the change in the potential energy. [Note that in this approximation, a linear function is being used for the difference between two inverse functions. $\Delta E = mgh$ for $\Delta E = -GmM$ $(1/(R + h) - 1/R)$.] (To 5 per cent, the approximation is good for $h \leq$ 200 miles.)

5. Find a parabola that approximates a circular well. Graph the two together to see over what range the parabola is a good approximation to the circle.

4.2 THE SINUSOIDAL FUNCTIONS—sin θ, cos θ, tan θ. The power functions $y = x^n$ describe variables that increase or decrease without limit. Many things in this world, however, are repetitious. We all go in circles. To describe such actions we must use mathematical functions whose values repeat periodically. One of the astonishing features of the sinusoidal functions is that they describe exactly, or closely approximate, a vast number of phenomena that are subject to oscillation or are dependent upon circular motion.

Consider the motion of a simple pendulum. Casual observation would provide the data points shown on the graph. As a first guess, one might

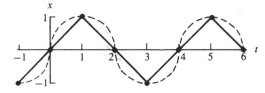 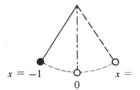

describe the motion by connecting the end-point data with straight lines, forming a sawtooth pattern. Such a model contains implications about the velocity and acceleration of the pendulum bob, as well as the position. First, the constant slopes indicate constant velocity ($v = \Delta x/\Delta t$) from end point to end point. At the turnaround, the math model calls for an abrupt change of velocity from positive to negative, requiring a quite unphysical infinite acceleration. Whether or not a pendulum or other oscillating object has constant velocity in midrange, it clearly must slow down in order to turn around. The sawtooth function is not a good model for such motion.

Since the pendulum orbit is part of a circular arc, perhaps the graph of $x(t)$ should be circular, as shown on the graph above. The math model would consist of semicircles describing $x(t)$. Such a function calls for a gradual slowing down with zero slope (and so, zero velocity) at the turnaround. Unfortunately, it also calls for infinite velocity as the bob goes through the equilibrium point. A circular function is clearly not a good model for oscillatory motion.

Plausibility arguments such as we have just made can rule out certain functions, but they cannot lead to the precise math model that does describe oscillatory motion. Indeed, it is not obvious that any simple function will be a good model, nor is it apparent that a function appropriate for a pendulum will also describe an object vibrating on a spring. As a matter of fact, the sinusoidal function that does describe the motion of a spring is only an approximation to pendulum motion.

The best-fit math model might be deduced from a more complete set of experimental data points. Alternatively, for a limited number of phenomena, a theory may link variables to form an equation with a simple exact solution. This latter is the case for many oscillatory systems, including an object bound by an ordinary spring. For both the spring and a pendulum restricted to small angles the solution is sinusoidal: $x = A \sin kt$. The displacement from equilibrium is x; the maximum value of x is the amplitude A; the constant, k, with dimensions of reciprocal time is necessary to make the product kt an angle; the time is t.

A. Graphs of the sinusoidal functions. Note the parity (even–odd) charac-

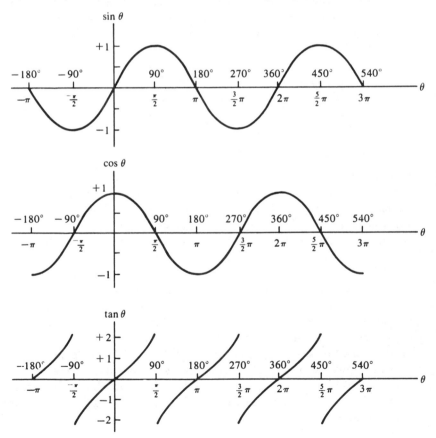

THE SIMPLE FUNCTIONS OF APPLIED MATH

teristics of the functions. The *sine* has odd parity: $\sin\theta = -\sin(-\theta)$. For instance, the sine of $30°$ is $+\frac{1}{2}$ and the sine of $-30°$ is $-\frac{1}{2}$. The *cosine* has even parity: $\cos\theta = +\cos(-\theta)$.

The cosine curve is basically the same as the sine curve, but is shifted $90°$ along the axis: $\cos\theta = \sin(\theta + 90°)$.

There are rules about finding sines and cosines for angles greater than $90°$. Rather than memorize such rules, just remember the graphs of the functions and figure out the position from 0 to $90°$ that corresponds to the angle in question. For instance, $\sin 480° = \sin(480° - 360°) = \sin(120°) = \sin(90° \pm 30°) = \sin 60°$.

B. The trigonometric source of the sinusoidal functions. The definitions of sine, cosine, and tangent are usually associated with the properties of right angles.

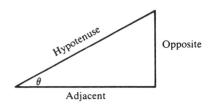

$$(4\text{--}12)$$

$$\sin\theta = \frac{\text{opp}}{\text{hyp}} \qquad \cos\theta = \frac{\text{adj}}{\text{hyp}} \qquad \tan\theta = \frac{\text{opp}}{\text{adj}} = \frac{\sin\theta}{\cos\theta}$$

$$\text{cosecant }\theta = \frac{\text{hyp}}{\text{opp}} \qquad \text{secant }\theta = \frac{\text{hyp}}{\text{adj}} \qquad \text{cotangent }\theta = \frac{\text{adj}}{\text{opp}} = \frac{\csc\theta}{\sec\theta}$$

C. The radian—the natural unit of angle. Circles have been divided into 360 equal parts since ancient times, probably because there are $365\frac{1}{4}$ days per year. The earth moves around the sun, or the sun moves along the ecliptic, about $1°$ per day. A choice of 360 parts rather than $365\frac{1}{4}$ allows many convenient fractional divisions.

For many purposes it is more convenient to define the unit of angle

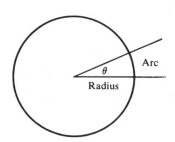

as follows: Inscribe a circle around the intersection of two lines. The measure of the angle between the two lines is defined to equal the ratio of arc length it subtends to radius:

$$\theta = \frac{\text{arc}}{r} \tag{4-13}$$

The ratio is independent of the size of the circumscribed circle. When arc length is equal to radius, the angle is the unit angle with measure 1.

Since the circumference of a circle is equal to $2\pi r$, there are 2π radians in a circle. In the diagram below, the circumference is divided into arc lengths equal to the radius. As you can see, there are approximately $6\frac{1}{3}$ such divisions.

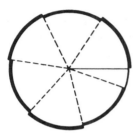

In one complete circle there are 2π rad or $360°$. Therefore, 2π rad = $360°$, or 1 rad = $57.3°$. Note that π rad = $180°$, and $\pi/2$ rad = $90°$.

The power-series approximation to the sinusoidal functions is described in Section 14.3. A first term in that approximation can be easily understood in terms of our diagram and definition of the radian. The angle is defined to

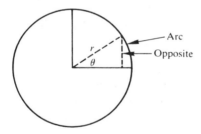

be $\theta = \text{arc}/r$. The sine of the angle is defined to be $\sin \theta = \text{opp/hyp} = \text{opp}/r$. For small angles, the arc is about the same length as the semichord labeled "opp," which is the opposite leg of a right triangle. Consequently,

$$\sin \theta \approx \theta, \qquad \text{for small } \theta \tag{4-14}$$

For example, for $\theta = 30° = 30/57.3$ rad = 0.522 rad, $\sin \theta = 0.50$. The approximation is good to about 2 parts in 50, or to about 4 per cent. For many purposes, the sine of a small angle can be figured in your head by using the extra approximation that 1 rad $\approx 60°$. For instance, $\sin 15° \approx \sin \frac{1}{4} \approx 0.25$. The actual value is 0.26.

D. The dimensionality of θ and other arguments. The sine, cosine, and tangent are functions of an angle, which is dimensionless. The measure of an angle is defined to be the ratio of two lengths. Nevertheless, we frequently want the argument of a sinusoidal function to consist of a time or a length. In the case of a pendulum, for example, the bob repeats its motion as a function of time. In this case, time can be considered in terms of fractions of the basic period of the motion. This fraction of the period is, in turn, the fraction of a complete cycle of 360° or 2π rad.

$$\sin\left(360°\frac{t}{T}\right) = \sin\left(2\pi\frac{t}{T}\right) = \sin\theta$$

The period, T, is the reciprocal of the frequency: $T = 1/f$. [0.1-s period $= 1/(10$ cycles/s).] The angular frequency in radians/second, ω, is related to the cyclical frequency, f: $\omega = 2\pi f$ (2π rad/s $= 1$ cycle/s). Consequently,

$$\sin\left(2\pi\frac{t}{T}\right) = \sin(2\pi ft) = \sin(\omega t) \tag{4-15}$$

The sine and cosine functions are simply 90° out of phase with each other. Time need not be measured so that the clock starts with $t = 0$ when $x = 0$ or maximum. In general there will be a phase angle at the beginning so that

$$x = A\sin(\omega t + \alpha) \tag{4-16}$$

$$\omega = 2\pi f = \frac{2\pi}{T}$$

$x = A\sin(\omega t + a)$

$\omega = 2\pi f \qquad f = \frac{2\pi}{T}$

In this case, $T = 4$ sec,

and $\omega = \frac{\pi}{2}$. The phase angle, a,

is equal to $\frac{\pi}{6}$ since at $t = 0$,

$x = 0.5A$. ∴ $0.5A = A\sin a$

and $a = 30° = \frac{\pi}{6}$

If the magnitude of a repeating variable is a function of length, the sinusoidal expression can be written in terms of fractions of the repetition length, or wavelength:

$$y = A\cos\left(2\pi\frac{x}{\lambda}\right) = A\sin\left(2\pi\frac{x}{\lambda} + \frac{\pi}{2}\right) \tag{4-17}$$

In this case, $\lambda = 4$, and

$$y = A\cos\left(\frac{\pi}{2}x\right) = A\sin\left(\frac{\pi}{2}x + \frac{\pi}{2}\right)$$

A traveling wave is a function of both x and t. Consider a pattern of ripples moving on the surface of a pond. A snapshot of the ripples showing their crests and troughs along one direction might be represented by $y = A \sin 2\pi(x/\lambda)$. The waves are moving past any particular point, however, and for that point the water height is rising and falling according to $y = A \sin 2\pi(t/T)$. If the ripples are moving in the positive x direction, the complete motion can be described as

$$y = A \sin 2\pi\left(\frac{t}{T} - \frac{x}{\lambda}\right) \tag{4-18}$$

As time increases positively, increasing the first half of the argument, x can increase positively so that the value of the total argument remains the same. Hence a particular point with constant amplitude (e.g., a crest) must be moving in the positive x direction.

The nondimensionality of the argument can be maintained through the following transformations, using the fact that velocity, frequency, and wavelength are related by $v = f\lambda$.

$$y = A \sin 2\pi\left(\frac{t}{T} - \frac{x}{\lambda}\right) = A \sin 2\pi f\left(t - \frac{x}{v}\right) = A \sin \omega\left(t - \frac{x}{v}\right) \tag{4-19}$$

In most real situations, the wave amplitude, A, is steadily decreasing as the wave progresses. Energy may be lost to friction and the wave may be spreading over a larger area, dissipating its intensity. The energy density of a wave is proportional to the square of its amplitude. (For a harmonic oscillator, $E_{\text{potential}} = \frac{1}{2}kx^2$; the total energy must equal $\frac{1}{2}kA^2$, where A is the maximum amplitude.) As a surface wave spreads out from a point in larger and larger circles, the energy is spreading out along the circumference. Since the length of the circumference increases as the first power of r ($C = 2\pi r$), the energy *density* must depend on $1/r$ and the amplitude on $1/\sqrt{r}$. The equation for such a wave spreading in two dimensions is

$$y = \frac{A}{\sqrt{r}} \sin \omega\left(t - \frac{r}{v}\right) \tag{4-20}$$

For a wave spreading in three dimensions, the energy density decreases as $1/r^2$, and so the amplitude would decrease as $1/r$. If there is also an absorption or decay with rate proportional to the amplitude, the wave solution for spherical waves would look like this:

$$y = \frac{e^{-(r \cdot a)}}{r} A \sin \omega\left(t - \frac{r}{v}\right) \tag{4-21}$$

E. The slopes, or derivatives, of sine and cosine. At the beginning of this section we presented a plausibility argument that neither a sawtooth nor a semicircular function could be a good model for repetitive motion of physical objects. The slopes of the graphs of these functions are related to

THE SIMPLE FUNCTIONS OF APPLIED MATH

the velocities of the motion and would require unphysical effects. We should now consider the velocity and acceleration connected with sinusoidal motion. If we plot $x(t)$, the slope (or first derivative) at any point is equal to the velocity at that time. If we plot the graph of these velocities, the slope of $v(t)$ is equal to the acceleration at that time: $v = dx/dt; a = dv/dt = d^2x/dt^2$.

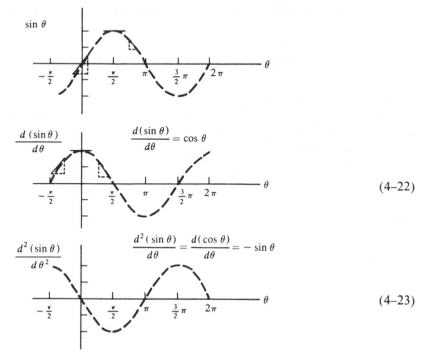

$$\frac{d(\sin\theta)}{d\theta} = \cos\theta \qquad (4\text{--}22)$$

$$\frac{d^2(\sin\theta)}{d\theta} = \frac{d(\cos\theta)}{d\theta} = -\sin\theta \qquad (4\text{--}23)$$

If $x = A \sin(\omega t + \alpha)$, then

$$v = \frac{dx}{dt} = \omega A \cos(\omega t + \alpha) \qquad (4\text{--}24)$$

$$a = \frac{d^2x}{dt^2} = \frac{dv}{dt} = -\omega^2 A \sin(\omega t + \alpha)$$

If $x(t)$ is sinusoidal, then $v(t)$ and $a(t)$ are also sinusoidal. This is plausible in terms of physical motion, and indeed agrees with experimental observation. With pendulum or spring oscillations, the maximum velocity occurs in the middle of the two end points, when the object is passing through the equilibrium position. When x, which is proportional to the sine, equals zero, v, which is proportional to the cosine, is maximum. At the end points where velocity slows to zero, the acceleration is maximum. Furthermore, the acceleration is always in the opposite direction from the displacement. When x is positive maximum, a is negative maximum. Note the important feature of the sine function: *it is the function whose second derivative is proportional to its negative self.*

4.2 THE SINUSOIDAL FUNCTIONS—sin θ, cos θ, tan θ

The radian is a "natural" unit of angle with respect to derivatives of the sinusoidal functions. The derivative, or slope, of the sine of an angle is *proportional* to the cosine of the angle. The proportionality constant is equal to 1 only if the angular unit is the radian: $(d/d\theta)(\sin\theta) = K\cos\theta$, and $K = 1$ if θ is in radians. The reason for this can be seen by observing the graph of $\sin\theta$ and remembering that $\sin\theta \approx \theta$ for small θ, if θ is expressed in radians. The slope at $\theta = 0$ is 1:

$$\left[\frac{d(\sin\theta)}{d\theta}\right]_{\theta\to 0} \approx \frac{d(\theta)}{d\theta} = 1$$

F. Circular motion and the sinusoidal functions. One reason for the pervasive appearance of the sinusoidal functions is that they represent one aspect of circular motion. From planets to bicycle wheels to electrons in the Bohr model, objects go in circles or in ellipses that are nearly circles. The circular motion itself is not sinusoidal, but the projection on an axis is. An

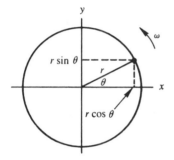

object on the circumference of a circle at angle θ has a projection on the x axis at $r\cos\theta$ and on the y axis at $r\sin\theta$. If the object is moving at constant velocity, ω, then $\theta = \omega t$. The projection on the x axis is $x = r\cos\omega t$. The projection on the y axis is $y = r\sin\omega t$.

Although the object moves with constant angular velocity around the circumference, the projections move back and forth along the axes, slowing down as they reach the rim and speeding up as they head for the center. The velocity of the projected point along the x axis is

$$v_x = \frac{dx}{dt} = -\omega r\sin\omega t \tag{4-25}$$

Note the reason for the minus sign. At $t = 0$, the x projection is at its maximum extension and during the following half-cycle moves to the left in the negative x direction. The velocity of the projected point along the y axis is $v_y = dy/dt = \omega r\cos\omega t$. At $t = 0$, the y projection point is passing through zero with maximum velocity, ωr, which is equal to v, the tangential velocity of the object moving on the circle.

THE SIMPLE FUNCTIONS OF APPLIED MATH

The acceleration of the x projection point is

$$a_x = \frac{d^2x}{dt^2} = -\omega^2 r \cos \omega t. \qquad (4\text{--}26)$$

At $t = 0$, when the point has its maximum positive value, its acceleration is maximum negative in the direction to restore the point to $x = 0$. The acceleration of the y projection point is $a_y = d^2y/dt^2 = -\omega^2 r \sin \omega t$. At $t = 0$, when the y projection point is passing through zero, its acceleration is zero.

This sinusoidal motion is called simple harmonic motion (SHM). It is simple compared with other oscillatory motion, and it is harmonic because, as we shall show in Section 14.4, the simple sine functions can be added together to form more complicated functions. In the sense of music chords, the simple functions can harmonize.

Projections on axes of circular motion are universal. The length of daylight is a sinusoidal function of annual time. Sea level is sinusoidal throughout each half-day, at least to first approximation. Any mechanism that turns rotary motion into linear drive, or vice versa, must involve SHM.

G. Vibrational motion and the sinusoidal functions. Simple harmonic motion appears in a multitude of phenomena that have nothing to do with circles. Most vibrating systems—springs, electrical oscillators, particles bound in potential wells—can be described, at least to first approximation, in terms of SHM. The necessary condition is that the system have an equilibrium condition, and that departures from that equilibrium be subject to a restoring acceleration proportional to the magnitude of the departure.

$$a_{\text{restoring}} = \frac{d^2x}{dt^2} = -kx \qquad (4\text{--}27)$$

The negative sign indicates that the restoring acceleration is in the opposite direction from the displacement from equilibrium, x.

The solution of the differential equation $d^2x/dt^2 = -kx$ requires a function $x(t)$ with the property that its second derivative is proportional to its negative self. That is just the property possessed by the sine function:

$$\frac{d^2(\sin \omega t)}{dt^2} = -\omega^2(\sin \omega t) \qquad (4\text{--}28)$$

Note that none of the power functions has such a property.

A prime example of a phenomenon satisfying the conditions for SHM is the behavior of a spring. Ordinary springs obey Hooke's law; i.e., $F_{\text{restoring}} = -kx$. The spring constant, k, is a function of the material and construction of the spring. Assuming that the mass of the spring is negligible compared with the bob at its end with mass, m,

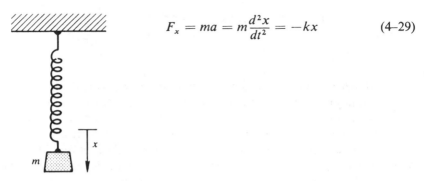

$$F_x = ma = m\frac{d^2x}{dt^2} = -kx \qquad (4\text{-}29)$$

The solution to the equation must be $x = A \sin(\omega t + \alpha)$. Differentiating this solution twice yields $d^2x/dt^2 = -\omega^2 A \sin(\omega t + \alpha)$. Substituting this value into Hooke's law,

$$m\frac{d^2x}{dt^2} = -m\omega^2 A \sin(\omega t + \alpha) = -kx = -kA \sin(\omega t + \alpha)$$

$$\omega^2 = 4\pi^2 f^2 = \frac{4\pi^2}{T^2} = \frac{k}{m} \qquad (4\text{-}30)$$

$$f = \frac{1}{2\pi}\sqrt{\frac{k}{m}}$$

The bob oscillates with SHM at a frequency that is inversely proportional to its mass.

An example of a motion that is approximately SHM is provided by the simple pendulum. The gravitational pull downward on the pendulum bob

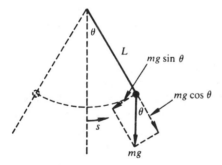

can be resolved into a component in the direction of the string and a component perpendicular to the string, and hence along the path of the bob. This latter is always in the direction to be a restoring force and is equal to

$$F_{\text{restoring}} = -mg \sin \theta \qquad (4\text{-}31)$$

Newton's second law for the motion is

$$F = ma = -mg \sin \theta = m\frac{d^2s}{dt^2} = mL\frac{d^2\theta}{dt^2} \qquad (4\text{-}32)$$

The arc length, s, is equal to $L\theta$, since by definition $\theta = s/L$. For positive s or θ, the restoring force is negative. The differential equation becomes

$$\frac{d^2\theta}{dt^2} = -\frac{g}{L}\sin\theta \qquad (4\text{-}33)$$

To satisfy this equation, we must find a function, $\theta(t)$, whose second derivative is proportional to the negative of the *sine of that function*. The sine function itself does not have that property, nor does any other simple function. A simple solution exists only if we make the approximation that $\sin\theta \approx \theta$ for small θ. Then the equation becomes

$$\frac{d^2\theta}{dt^2} \approx -\frac{g}{L}\theta \qquad (4\text{-}34)$$

Its solution is

$$\theta = \Theta_{max}\sin(\omega t + \alpha) \qquad \text{with } \omega = \sqrt{\frac{g}{L}} \qquad (4\text{-}35)$$

The period given by this expression is good to within 1 per cent up to $\Theta_{max} = 23°$ and within 4 per cent up to $\Theta_{max} = 45°$.

An example of an electrical system that oscillates is the inductance–capacitance "tank" circuit. Adding the potential changes around the circuit,

$$L\frac{di}{dt} + \frac{q}{C} = 0 \qquad (4\text{-}36)$$

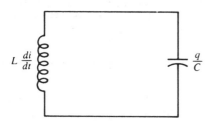

Since the current i is related to the charge q by $i = dq/dt$,

$$L\frac{d^2q}{dt^2} + \frac{q}{C} = 0 \qquad \text{or} \qquad \frac{d^2q}{dt^2} = -\frac{1}{LC}q \qquad (4\text{-}37)$$

Once again, this equation is satisfied by a sine function, since the second derivative must be proportional to the negative function.

$$q = Q\sin(\omega t + \alpha) \qquad \text{with } \omega = \sqrt{\frac{1}{LC}} \qquad \text{or} \qquad f = \frac{1}{2\pi}\sqrt{\frac{1}{LC}} \qquad (4\text{-}38)$$

Note that in all these cases the negative sign in the equation is not a trivial matter. Without it the equation would describe a situation in which the acceleration term is in the same direction as the displacement from equilibrium. Rather than an oscillation, an avalanching effect would be produced. In the next section we show that the exponential function has the required properties to describe such a situation.

Yet another reason for the prevalence of SHM is described on p. 33 in the section on the quadratic function. As explained there, objects bound in potential wells of almost any kind are subject to a restoring force proportional to the negative of their displacement from equilibrium, at least for small displacements. Atoms in crystals, for instance, vibrate with simple harmonic motion.

H. Trig identities and relationships.

1. For any angle θ, $\underline{\sin^2 \theta + \cos^2 \theta = 1}$. (4–39)

   This relationship follows from the definitions of sine and cosine, and the Pythagorean theorem: $c^2 = a^2 + b^2$.

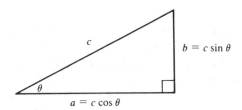

2. The sum of the interior angles of every triangle is 180°.

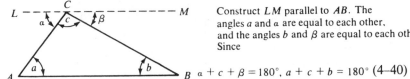

Construct LM parallel to AB. The angles a and α are equal to each other, and the angles b and β are equal to each other. Since

$$\alpha + c + \beta = 180°, \quad a + c + b = 180° \quad (4\text{–}40)$$

3. Any triangle circumscribed by a circle, with one side the diameter of the circle, is a right triangle with the diameter of the circle being the hypotenuse.

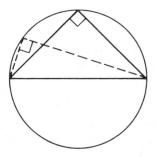
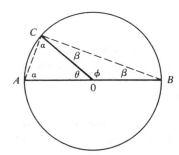

   For any circumscribed triangle, ACB, draw the radius CO. The two triangles AOC and COB are each isosceles with sides equal to the radius. Therefore,

THE SIMPLE FUNCTIONS OF APPLIED MATH

$$2\alpha + \theta = 180° \qquad\qquad 2\beta + \phi = 180° \qquad \theta + \phi = 180°$$
$$2\alpha + \theta + 2\beta + \phi = 360° \qquad\qquad (4\text{-}41)$$
$$2\alpha + 2\beta = 180°$$
$$\alpha + \beta = 90°$$

4. Sine and cosine values for common angles:

$$\sin 0° = \cos 90° = 0$$

$$\sin 30° = \cos 60° = \tfrac{1}{2} = 0.500$$

$$\sin 45° = \cos 45° = \frac{1}{\sqrt{2}} = 0.707$$

$$\sin 60° = \cos 30° = \frac{\sqrt{3}}{2} = 0.866$$

$$\sin 90° = \cos 0° = 1$$

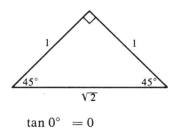

In the 45° isosceles triangle, the hypotenuse has a length of $\sqrt{2}$ if each leg = 1, from ($c^2 = a^2 + b^2$)

$$\therefore \sin 45° = \sqrt{\tfrac{1}{2}}$$

$$\tan 0° = 0$$
$$\tan 45° = 1$$
$$\tan 90° = \infty$$

In the 60° equilateral triangle, a bisector divides the base in half as shown. The length of the bisector is:

$$\sqrt{(2)^2 - (1)^2} = \sqrt{3}$$

$$\therefore \sin 60° = \frac{\sqrt{3}}{2}$$

5.
$$\sin(\theta \pm \phi) = \sin\theta\cos\phi \pm \cos\theta\sin\phi \qquad \sin 2\theta = 2\sin\theta\cos\theta$$
$$\cos(\theta \pm \phi) = \cos\theta\cos\phi \mp \sin\theta\sin\phi \qquad \cos 2\theta = \cos^2\theta - \sin^2\theta$$
$$= 1 - 2\sin^2\theta$$
$$= 2\cos^2\theta - 1$$

$$\tan(\theta \pm \phi) = \frac{\tan\theta \pm \tan\phi}{1 \mp \tan\theta\tan\phi} \qquad\qquad \tan 2\theta = \frac{2\tan\theta}{1 - \tan^2\theta}$$

$$\sin \tfrac{1}{2}\theta = \pm\sqrt{\frac{1 - \cos\theta}{2}}$$

$$\cos \tfrac{1}{2}\theta = \pm\sqrt{\frac{1 + \cos\theta}{2}}$$

$$\tan \tfrac{1}{2}\theta = \sqrt{\frac{1 - \cos\theta}{1 + \cos\theta}} = \frac{\sin\theta}{1 + \cos\theta} = \frac{1 - \cos\theta}{\sin\theta}$$

6. $\sin\theta + \sin\phi = 2\sin\tfrac{1}{2}(\theta + \phi)\cos\tfrac{1}{2}(\theta - \phi)$
   $\sin\theta - \sin\phi = 2\cos\tfrac{1}{2}(\theta + \phi)\sin\tfrac{1}{2}(\theta - \phi)$
   $\cos\theta + \cos\phi = 2\cos\tfrac{1}{2}(\theta + \phi)\cos\tfrac{1}{2}(\theta - \phi)$
   $\cos\theta - \cos\phi = -2\sin\tfrac{1}{2}(\theta + \phi)\sin\tfrac{1}{2}(\theta - \phi)$

7. $\sin\theta \sin\phi = \tfrac{1}{2}\cos(\theta - \phi) - \tfrac{1}{2}\cos(\theta + \phi)$
   $\cos\theta \cos\phi = \tfrac{1}{2}\cos(\theta - \phi) + \tfrac{1}{2}\cos(\theta + \phi)$
   $\sin\theta \cos\phi = \tfrac{1}{2}\sin(\theta + \phi) + \tfrac{1}{2}\sin(\theta - \phi)$

8. Law of sines:
$$\frac{a}{\sin A} = \frac{b}{\sin B} = \frac{c}{\sin C}$$

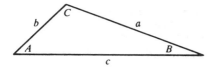

9. Law of cosines:
$$c^2 = a^2 + b^2 - 2ab\cos C$$
$$b^2 = a^2 + c^2 - 2ac\cos B$$
$$a^2 = b^2 + c^2 - 2bc\cos A$$

10. $\sin\theta = \dfrac{e^{i\theta} - e^{-i\theta}}{2i}$ $\qquad e^{i\theta} = \cos\theta + i\sin\theta$

$\cos\theta = \dfrac{e^{i\theta} + e^{-i\theta}}{2}$ $\qquad e^{-i\theta} = \cos\theta - i\sin\theta$

The plausibility of these identities involving the sinusoidal functions and the imaginary exponentials is best seen by expanding each side in terms of power series. (See Section 14.3.) Using these relationships, you can easily demonstrate the trig identities listed under 5, 6, and 7 of this section. For instance,

$$\sin(\theta + \phi) = \frac{e^{i(\theta + \phi)} - e^{-i(\theta + \phi)}}{2i} = \frac{e^{i\theta}e^{i\phi} - e^{-i\theta}e^{-i\phi}}{2i}$$

$$= \frac{1}{2i}[(\cos\theta + i\sin\theta)(\cos\phi + i\sin\phi)$$

$$-(\cos\theta - i\sin\theta)(\cos\phi - i\sin\phi)]$$

$$= \frac{1}{2i}(\cos\theta \cos\phi + i\sin\theta \cos\phi + i\cos\theta \sin\phi$$

$$-\sin\theta \sin\phi - \cos\theta \cos\phi + i\sin\theta \cos\phi$$

$$+ i\cos\theta \sin\phi + \sin\theta \sin\phi)$$

$$= \frac{1}{2i}(2i \sin \theta \cos \phi + 2i \cos \theta \sin \phi)$$

$$= \sin \theta \cos \phi + \cos \theta \sin \phi$$

SAMPLES

1. On good graph paper, plot $\sin \theta$ versus θ from 0 to 2π with careful detail between 0 and $\pi/2$. Use the top third of the graph paper for this plot. In the middle third of the paper plot the slopes of the sine graph above. Measure these slopes carefully with ruler and sharp pencil, especially during the first quarter-cycle. Note that with θ expressed in radians, not degrees, the maximum slope is equal to 1. In the bottom third of the paper, plot the slopes of the middle graph as measured with ruler and pencil.

2. Use the graphs, small angle approximation, and memorized values for 0, 30, 45, 60, and 90° angles to find approximate values for (a) $\sin 5°$, (b) $\cos 190°$, (c) $\tan 315°$, (d) $\sin 210°$, (e) $\tan 30°$, (f) $\cos 750°$, and (g) $\sin -120°$.

3. The moon is 2000 miles in diameter and about 240,000 miles from earth. What is the angular width of the moon as seen from the earth in radians and in degrees?   (~ 0.008 rad or $\sim\frac{1}{2}°$.) What is the angular width of your thumbnail when held at arm's length?

4. What is the angular velocity of the earth around its axis in radians/second?   (7.3×10^{-5} rad/s)

5. If an object is constrained to move in a circular path, it must be subject to a centripetal force producing a centripetal acceleration $= -v^2/r = -\omega^2 r$. The x projection of the object is $x = r \cos \omega t$; the y projection is $y = r \sin \omega t$. Find the x and y projections of acceleration, and then show that the vector sum of the accelerations at any angle is the centripetal acceleration given above.

6. Set up the differential equation for an object with mass m that is bound by a spring with constant k. Find an appropriate solution if $t = 0$ when the object is at the top of its vibration. Find the period if $m = 0.2$ kg and $k = 20$ newtons (N)/m.   ($y = A \sin 2\pi[(t/0.63) + \frac{1}{4}]$, $T = 0.63$ s.)

7. Set up a simple pendulum so that it swings from one small region (such as a knot) attached to a solid support. With a protractor grid in the background, time 5 to 10 full swings at each starting angle in order to get 1 per cent precision. Measure and graph T as a function of the starting angle θ. Find the deviation from the simple formula for the period for angles between 30 and 60°.

8. Use the series expansions for e^x and for $\sin x$ to show that $\sin x = (e^{ix} - e^{-ix})/2i$. Use the latter relationships to show that $\sin \theta \sin \phi = \frac{1}{2} \cos(\theta - \phi) - \frac{1}{2} \cos(\theta + \phi)$.

4.2 THE SINUSOIDAL FUNCTIONS—$\sin \theta$, $\cos \theta$, $\tan \theta$

4.3 THE EXPONENTIAL FUNCTION: $y = a^x$. In ancient times, so the story goes, a king in India received sage advice from a sage. Wishing to reward the old man, the king offered him any prize he wished. "Sire," said the sage, "I am a humble man of modest desires. Give me for my labors only grains of rice to match the squares on a chessboard. One grain for the first square, two for the second, four for the third, and so on, doubling the number for each successive square." The king, being no mathematician, was embarrassed that his councilor would be satisfied with so little. You, on the other hand, realize that by asking for $\sum_{x=0}^{63} 2^x$ grains, the greedy old man was demanding more rice than has ever grown. This section is intended to remind you of the properties and peculiar growing power of the exponential function.

A. The nature of the exponential. Note first the vast difference between a^x and x^a; the first is the exponential, the second is a power function. Although the contrast between 2^{63} and 63^2 is dramatic, no specific values of functions are pertinent in comparing functional characteristics. Consider instead the graphs of 2^x and x^2. The exponential will eventually outrun any power function.

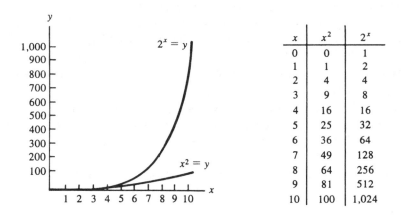

x	x^2	2^x
0	0	1
1	1	2
2	4	4
3	9	8
4	16	16
5	25	32
6	36	64
7	49	128
8	64	256
9	81	512
10	100	1,024

The significance of a^{+x} (e.g., 2^4, 10^3) is plain enough. The variable x can take on negative values as well: $a^{-x} = 1/a^x$ (e.g., $10^{-3} = 0.001$). Products of exponentials follow this rule: $a^x \times a^y = a^{x+y}$. For instance,

$$2^2 \times 2^3 = 2^5 \quad \text{and} \quad 10^2 \times 10^{-3} = 10^{-1}$$
$$(4 \times 8 = 32) \quad\quad (100 \times 0.001 = 0.1)$$

This rule defines the significance of a^0, since

$$a^{+x} \times a^{-x} = a^0 \quad \text{and} \quad a^{+x} \times a^{-x} = a^x \times \frac{1}{a^x} = 1$$

Hence, $a^0 = 1$ for any value of the base, a.

B. The base of the exponential.   In the following diagram we show the exponential function for three different bases. Two of them, 2^x and 10^x, are obviously useful in describing many phenomena of interest. If pairs continually double, the number increases as 2^x. Furthermore, this function describes the place values in the binary counting system: 1, 2, 4, 8, 16, etc. Since our main counting system is decimal, 10^x describes the customary place values: 1, 10, 100, 1000, etc.

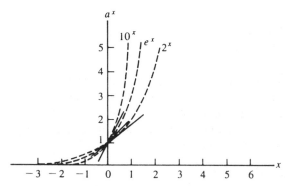

The third curve on the graph represents the function e^x. This particular number, e, is called the natural base and has the value (approximately) $e = 2.718$. What is so natural about the natural base? Inspect the three curves and note that necessarily they all go through the point $(0, 1)$. Their slopes at that point are indicated. The slope of e^x is equal to 1. The graph of the slope, or in other words, the graph of the first derivative of e^x, is shown in the next diagram.

As you can check for yourself,

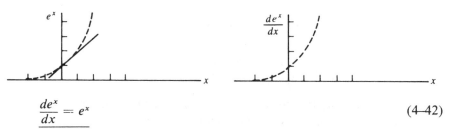

$$\frac{de^x}{dx} = e^x \qquad\qquad (4\text{--}42)$$

Hence

$$\left(\frac{de^x}{dx}\right)_{x=0} = e^0 = 1$$

The function e^x has a derivative *equal* to itself. The derivative for the exponential to any other base is *proportional* to itself: $da^x/dx = a^x \ln a$. The natural logarithm, $\ln a$, is described in the next section. The values for the two other cases of interest are $d2^x/dx = (0.69)2^x$ and $(d10^x/dx) = (2.3)10^x$. The general formula for the derivative of the exponential is

4.3 THE EXPONENTIAL FUNCTION: $y = a^x$

$$\frac{d}{dx}a^u = \ln a \; a^u \frac{du}{dx} \qquad (4\text{-}43)$$

For instance,

$$\frac{d}{dx}2^{x^2} = (0.69)2x2^{x^2}$$

$$\frac{d}{dx}e^{-kx} = -ke^{-kx}$$

$$\frac{d}{dx}e^{-x^2/\sigma^2} = -\frac{2x}{\sigma^2}e^{-x^2/\sigma^2}$$

C. Transforming bases. A general method for transforming bases of exponentials is useful. We must appeal to the relationship between logarithms and exponentials that will be demonstrated in the next section. First we assert that an exponential to one base can be changed to another by simply multiplying the exponent by a constant:

$$a^x = b^{kx} \qquad (\text{e.g., } 4^x = 2^{2x}) \qquad (4\text{-}44)$$

To find the constant, k, take the logarithm of both sides to the base b:

$$x \log_b a = kx \log_b b = kx$$

Therefore,

$$k = \log_b a$$

and

$$\underline{a^x = b^{x \; \log_b a}} \qquad (4\text{-}45)$$

For instance,

$$e^x = 10^{0.434x} \qquad (\log_{10} e = 0.434)$$
$$10^x = e^{2.3x} \qquad (\log_e 10 = 2.303)$$
$$2^x = e^{0.69x} \qquad (\log_e 2 = 0.693)$$
$$2^x = 10^{0.30x} \qquad (\log_{10} 2 = 0.301)$$

There is an easy way to remember a close approximation for changing large values of e^x and 2^x to the equivalent numbers of 10^y:

$$\underline{2^{10} \approx 10^3} \qquad \text{since } 2^5 = 32, \; 2^{10} = (32)^2 = 1024 \approx 1000$$
$$(\text{e.g., } 2^{30} \approx 10^9; \; 2^{16} \approx 6 \times 10^4)$$
$$\underline{e^3 = 20.09} \qquad \text{then: } e^{15} \approx 20^5 = 32 \times 10^5$$

One particular change of exponential base that is frequently employed makes use of an approximation, although this feature is seldom mentioned. If a quantity is increasing by a fixed small percentage per unit time, then the magnitude at time t is given by

$$p = P(1 + r)^{t/T} \qquad (4\text{-}46)$$

where P is the amount when $t = 0$, and r is the fractional increase per time T. For instance, P might be a sum of money left to accumulate compound interest for t years at an interest rate of 0.06 per year, T. If the increase in the unit time T is very small, the magnitude is *approximately* given by $p = Pe^{rt/T}$. To see why this is so, change the original base in the formal way:

$$(1 + r)^{t/T} = e^{t/T \ln(1+r)} \qquad (4\text{-}47)$$

The series approximation for the natural log is (see Section 14.3, p. 195) $\ln(1 + r) = r - \frac{1}{2}r^2 + \frac{1}{3}r^3 - \frac{1}{4}r^4 + \cdots$. Thus,

$$(1 + r)^{t,T} \approx e^{rt,T} \qquad (4\text{-}48)$$

The difference in growth of capital compounded continuously compared with annual compounding is tabulated below. The approximation to the base e corresponds to letting r and T get smaller and smaller, maintaining the ratio r/T. For instance, 6 per cent interest per year becomes 0.5 per cent interest per month. For continuous compounding, $r \to 0$, but r/T still remains 6 per cent per year.

Year	6%/yr Continuous $e^{rt,T}$	6%/yr Annual $(1 + r)^{t/T}$
0	1000	1000
1	1062	1060
2	1127	1123.5
3	1197	1191.0
4	1271	1262.5
5	1350	1338.0

D. Uses of the exponential function.

1. Compound interest growth. The actual formula for compound interest growth at finite intervals is, as we have already seen, $p = P(1 + r)^{t,T}$. The principal p at time t is equal to the original principal P compounded with the interest rate r per time interval T. For instance, the principal at the end of the third year with an original investment of $1000 at an interest rate of 6 per cent per year is equal to $p = \$1000(1 + 0.06)^3 = \$1191$.

Another way of deriving the approximate solution to the base e is this: The increase of principal Δp in a short time interval Δt is proportional to the principal p at the start of the interval, and to the interest rate r, and also to the length of the interval Δt. $\Delta p = pr \, \Delta t$. According to this formula, if you plot p versus t, the slope is proportional to p and the proportionality constant is r. What function is it whose slope is proportional to itself? The exponential has this property. A solution to the equation is $p = Pe^{rt}$, where t is measured in the same units used to express r, usually fraction per *year*. The derivation is based on continuous compounding because in the original statement it was assumed that the increase is proportional to the time interval and to the principal at a particular time. These conditions are met only in the limit as $\Delta t \rightarrow 0$.

2. Mortgage payments. In making monthly mortgage payments, you pay the same amount each month for a predetermined number of years. Part of the monthly payment is for the interest owed for 1 month on the remaining debt and the rest goes to pay off the principal. If you wanted to borrow $10,000 for 20 years at 6 per cent interest (if you could be so lucky), how much would you have to pay each month?

principal repayment + month's interest = constant monthly payment
$$-\Delta p \qquad + \qquad pr\,\Delta t \qquad = \qquad K\,\Delta t$$

The minus sign in front of the principal repayment indicates that there is a decrease in the principal. The resulting differential equation is

$$\frac{dp}{dt} = pr - K \tag{4-49}$$

A simple exponential cannot be a solution to this equation, since here the first derivative of the function is proportional to the function minus a constant. Sometimes a solution to a differential equation can be manufactured by examining the plausible behavior of the function. In this case, if we plot p versus t, we know that $p = P$ when $t = 0$, and $p = 0$ when $t = T$, where T is the total number of payments. Furthermore, p decreases very slowly at first, since for large p, $(pr - K)$ must be nearly zero. When t is large, p will be small and the monthly payment K goes mostly toward reducing p. The graph of the function must look something like the diagram. It seems reason-

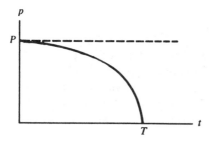

able to expect that the curve is in some way exponential. To make the function equal to zero when $t = T$, the numerator might contain $e^{rT} - e^{rt}$. To make this term equal to unity when $t = 0$, there must be a term in the denominator equal to $e^{rT} - 1$. A plausible final product then is

$$p = P\frac{e^{rT} - e^{rt}}{e^{rT} - 1} \tag{4-50}$$

The first derivative of this function is

THE SIMPLE FUNCTIONS OF APPLIED MATH

$$\frac{dp}{dt} = \frac{-rP}{e^{rT}-1}e^{rt}$$

Substituting back into equation (4–49),

$$\frac{-rP}{e^{rT}-1}e^{rt} = rP\frac{e^{rT}-e^{rt}}{e^{rT}-1} - K$$

$$K = rP\frac{e^{rT}}{e^{rT}-1} = rP\frac{1}{1-e^{-rT}}$$

Since an appropriate constant can be found, and since this function satisfies the equation, it must be the solution. Using the numbers from our original example, the monthly payment would be

$$K = [0.005(\text{per month})]\$10,000\frac{1}{1-e^{-0.005(240)}} = \$50\frac{1}{1-e^{-1.2}}$$
$$= \$71.50$$

The principal remaining at any time is

$$P = \$10,000\left(\frac{3.32 - e^{0.005t}}{2.32}\right)$$

At the halfway mark, when $t = 120$ months, $p = \$6460$. In making this calculation we have assumed the approximation that payments are continuous. Bankers, of course, don't fool around with approximations or with calculus either. They have standard tables in which to find such numbers. According to one of these tables, the monthly payment for our particular example should be \$71.65.

3. Radioactive decay. Here is another prime example of how one can use plausibility arguments to find a math model for a physical phenomenon. If you start out with a lot of radioactive atoms, you know that they will decay, transforming themselves into other atoms, so that after a while there will be only a few of the original kind left. The end points of a possible curve are indicated on the sketch; the question is, which curve represents the function?

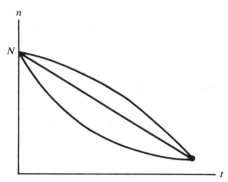

The number decaying, Δn, per interval of time, Δt, must be proportional

4.3 THE EXPONENTIAL FUNCTION: $y = a^x$

to the number there at that time, n, and to the length of the time interval, Δt.

$$\Delta n = -\lambda n \Delta t \tag{4-51}$$

The proportionality constant is λ, and the negative sign is necessary because Δn must be negative if n is decreasing. As for the reasonableness of the assumed proportionalities, you would expect twice as many decays from 2 million atoms as from 1 million, and you would expect twice as many decays in 2 s as from 1 s of counting, assuming that the decay time for an average atom is much greater than 1 s.

The resulting differential equation is $dn/dt = -\lambda n$. This equation is satisfied by a function whose first derivative with respect to time is proportional to itself; the function having that property is the exponential.

$$n = Ne^{-\lambda t} \tag{4-52}$$

When $t = 0$, $n = N$. The function satisfies the equation, as you can see by differentiating it and substituting it back in. Evidently, the graph of $n(t)$ must be represented by the lower curve in the sketch.

For every interval of time $t = 1/\lambda$, the number of atoms left is $1/e$, or 0.367, of the number at the beginning of the interval. Frequently, it is convenient or customary to describe the decay in terms of half-lives, the time it takes for half the atoms to decay. This corresponds to using the base 2 for the exponential instead of the base e. The transformation follows, with $T = $ half-life:

$$n = N2^{-(t/T)} = Ne^{-(t/T)\ln 2} = Ne^{-(0.69/T)t} \tag{4-53}$$

The half-life and proportionality constant are thus related by

$$T = \frac{0.69}{\lambda} \tag{4-54}$$

In some introductory physics courses, radioactive decay is simulated by a special dice game. Start with a large number of dice, throw them and remove all of any particular number—say, the 3's. Continue doing this, casting only the remaining dice, and graph the number that remain as a function of the number of throws. The larger the number of dice originally present, the closer will be the decay curve to an exponential. The math model for this process is usually thought to be

$$\Delta n = -\lambda n \, \Delta t = -\tfrac{1}{6} n \, \Delta t \qquad (\Delta t = 1, \text{ the interval for each throw})$$

Taking the next step and assuming that a differential equation can be formed is not quite correct, however, as careful data gathering will show. The decay constant is large in terms of the steps Δt. The correct expression (see p. 75) is

$$n = N(1 - r)^t = N(\tfrac{5}{6})^t$$

where $r = \tfrac{1}{6}$ is the probability of getting any particular number on the die.

THE SIMPLE FUNCTIONS OF APPLIED MATH

To turn this into an exponential with the base e, we use the standard transformation:

$$n = N(\tfrac{5}{6})^t = Ne^{t \ln (5/6)} = Ne^{-t \ln (6/5)} = Ne^{-0.182t}$$

Compare this accurate expression with the approximation when it is assumed that decay is continual with a constant of $\frac{1}{6}$, which is equal to 0.167:

$$n = Ne^{-0.167t}$$

The "half-life" in terms of number of throws is $T = 0.69/0.182 = 3.8$ and not $0.69/0.167 = 4.1$. The difference between these two half-lives is readily apparent from a semilog graph of the game, starting with as few as 100 dice.

For information on graphing exponential functions on semilog paper, consult Section 3.H, p. 25. The standard technique for finding the exponential decay constant is to measure the slope of the experimental curve on a semilog plot.

SAMPLES

1.   If you leave $100 in a bank with interest at 5 per cent to be compounded *annually*, how much will you have in 20 years?   ($265) How much would you have if the compounding took place *continually*?   ($272)

2.   With the mortgage example given on p. 58, find the time when the mortgage is half paid off: $p = P/2$.   (13 yrs)   For the terms of this mortgage, find the total amount paid at maturity with interest rates at 5, 6, and 7 per cent.

3.   Do the experiment with the dice described on p. 60. Graph the data and find the decay constant.

4.   On the top half of a sheet of graph paper, plot $y = 2^x$ with sufficient precision so that you can measure accurately the slopes at various points. Plot these values of the slopes directly below on the bottom half of the paper. Show graphically that $d(2^x)/dx = 0.69(2^x)$.

5.   The error involved in using the expression for continual compounding instead of annual compounding is. $\Delta p = Pe^{rt} - Pe^{(r+\Delta r)t}$, where Δr represents the terms ignored in the expansion of $\ln(1 + r)$: $\Delta r = -\frac{1}{2}r^2 + \frac{1}{3}r^3 - \frac{1}{4}r^4 + \cdots$. Using only the term containing r^2, find the fractional error $\Delta p/p$ and see if this agrees with the tabular data differences on p. 57.

6.   Do a Fermi calculation to check the assertion that the rice demanded by the Indian sage (see p. 54) is more than all the rice that has ever grown. Average world production of rice in the 1960s was 2×10^8 tons per year. Milled rice has a mass of about 20 mg per grain.

7.   Show that if all the gasoline tax is applied to the building of new roads, then the number of miles of highway should increase exponentially, if the number of miles driven is proportional to the number of miles of road in existence.

4.3 THE EXPONENTIAL FUNCTION: $y = a^x$

4.4 THE LOGARITHMIC FUNCTION: $y = \log_b x$. Long ago, it is reported, the Lord was walking in the Garden at noonday. He saw that the lions had cubs, the wolves had pups, the hens had chicks, and He was pleased. Then He saw two snakes all by themselves. "Why are there no little snakes?" he asked. "Did I not command you to be fruitful and multiply?" "We cannot," they said, "for we are adders." Some time thereafter, the Lord walked again in the Garden and found that the snakes were wrapped around some sticks. Lo, they had with them many baby snakes. "How were you able to multiply?" said the Lord. "I thought you were only adders." "Yes," replied the snakes, "but now we have logs."

Ever since that time the schools have taught that the purpose of logarithms is to aid in computation involving long multiplications. Actually, not even the U.S. Navy uses logs for that purpose any more. It is true that the logarithm system was invented by Napier almost 400 years ago to simplify calculations. Nowadays anyone who has an arithmetic problem in navigation or astrology or similar technical work uses an electronic computer. If only three significant figures are required in multiplication or division, you can use a slide rule, which is, as we shall see, just a mechanical log table. Nevertheless, the logarithm, as a function, has vitally unique features and is a model for many phenomena.

A. Definition of the logarithm.

x	0.001	0.01	0.1	1	10	100	1000
x	10^{-3}	10^{-2}	10^{-1}	10^0	10^1	10^2	10^3
$\log_{10} x$	-3	-2	-1	0	1	2	3

In this array we have written a row of numbers, x, that increase steadily by a factor of 10. The middle row simply indicates the exponential way of writing these numbers; it is a row of the same quantities, x. In the bottom row, rather than use the clumsy notation of 10^y, we write only the exponent itself and *define* these numbers to be the logarithms of x to the base 10. If $x = 10^y$, then by definition $y = \log_{10} x$.

Note what happens if we multiply any of the x numbers together:

$$0.1 \times 100 = 10$$
$$10^{-1} \times 10^2 = 10^{(2-1)} = 10^1$$

In multiplying two exponentials with the same base, you simply add the exponents. But the exponents are what we have defined to be logs. Hence

$$\log_{10}10^{-1} + \log_{10}10^2 = \log_{10}(10^{-1} \times 10^2) = \log_{10}10^1$$
$$-1 \quad + \quad 2 \quad = 1$$

To multiply any of the numbers, x, together, add their logs and find the new number corresponding to that sum log. In the example above, the sum log is equal to 1, which is the log of 10.

All this would be a conceit if we were to deal only with powers of 10. What we need is a set of logs corresponding to any number. Working in the range between 1 and 10, we can fill in other numbers in the following straightforward though laborious way. Find the square root of 10, or in the exponential form, $10^{1/2}$. It is 3.17. (For a rational method of taking square roots, see p. 191.) On the expanded section of the number rows, we have entered 3.17 in x, and 0.50 in the logs. The process can be continued by taking the cube root, the fourth root, the sixth root (the square root of the cube root), the five-sixth root (by multiplying the square root and the cube root), etc.

$$\sqrt[3]{10} = 10^{1/3} = 2.15 \qquad\qquad \log_{10} 2.15 = \tfrac{1}{3} = 0.333$$
$$\sqrt[4]{10} = 10^{1/4} = 1.78 \qquad\qquad \log_{10} 1.78 = \tfrac{1}{4} = 0.250$$
$$\sqrt[6]{10} = 10^{1/6} = (10^{1/3})^{1/2} = 1.47 \qquad \log_{10} 1.47 = \tfrac{1}{6} = 0.167$$
$$\sqrt[6]{10} = 10^{5/6} = (10^{1/3})(10^{1/2}) = 6.82 \qquad \log_{10} 6.82 = \tfrac{5}{6} = 0.832$$
$$\sqrt[3]{10} = 10^{2/3} = (10^{1/6})(10^{1/2}) = 4.66 \qquad \log_{10} 4.66 = \tfrac{2}{3} = 0.667$$
$$\sqrt[12]{10} = 10^{11/12} = (10^{2/3})(10^{1/4}) = 8.35 \qquad \log_{10} 8.35 = \tfrac{11}{12} = 0.922$$

The expanded array in the range 1 to 10 is shown below:

x	1	1.47	1.78	2.15	3.17	4.66	6.82	8.35	10
x	10^0	$10^{1/6}$	$10^{1/4}$	$10^{1/3}$	$10^{1/2}$	$10^{2/3}$	$10^{5/6}$	$10^{11/12}$	10^1
$\log_{10} x$	0	0.167	0.250	0.333	0.500	0.667	0.832	0.922	1.000

These tabular data are graphed below. A working table of logs to the base 10 appears on p. 248.

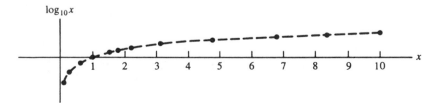

In finding and plotting the logarithms for the interval between 1 and 10, we have really found the logarithms for any decade interval. The logarithm of $21.5 = \log_{10} (2.15 \times 10^1) = \log_{10} 2.15 + \log_{10} 10^1 = 0.333 + 1 = 1.333$. The logarithm of 215 equals 2.333. The logarithm of 0.215 is *not* -1.333. Instead, $\log_{10} 0.215 = \log_{10} (2.15 \times 10^{-1}) = \log_{10} 2.15 + \log_{10} 10^{-1} = 0.333 - 1.000 = -0.667$. Several of these negative logs are also plotted on the graph.

There is a scheme sometimes taught in the schools for dealing with the logs of numbers smaller than 1 by adding and subtracting 10 as follows. They would call the log of 0.215, $9.333 - 10$. It is a peculiar system and should be forgotten. In finding logs, always express the number in terms of a number between 1 and 10, multiplied by a power of 10 (e.g., $5832 = 5.832 \times 10^3$; $0.062 = 6.2 \times 10^{-2}$). This system is both foolproof and rational.

We saw that with the integral powers of 10 you can multiply two numbers by adding their exponents, which are defined as logs. The logs for all numbers have this property of being additive for multiplication of the arguments. Here are the rules, each with an example:

1. $\log (ab) = \log a + \log b$
   $\log (2 \times 5) = \log 2 + \log 5 = 0.3010 + 0.6990 = 1.0000$
   $\qquad = \log 10$

2. $\log (a/b) = \log a - \log b$
   $\log \frac{8}{4} = \log 8 - \log 4 = 0.9031 - 0.6021 = 0.3010$
   $\qquad = \log 2$

   (4-55)

3. $\log a^n = n \log a$
   $\log 2^3 = 3 \log 2 = 3(0.3010) = 0.9030 = \log 8$

4. $\log \sqrt[n]{a} = \log a^{1/n} = 1/n \log a$
   $\log \sqrt[3]{2} = \log 2^{1/3} = \frac{1}{3} \log 2 = \frac{1}{3}(0.3010) = 0.1003$
   $\qquad = \log 1.26$

B. Logs to other bases—the natural log. Our choice of powers of 10 was quite arbitrary. Here are number arrays corresponding to two other bases of interest. We have not filled in the detailed numbers of the prime interval, but it could be done using the same method of roots that we employed before.

x	$\frac{1}{8}$	$\frac{1}{4}$	$\frac{1}{2}$	1	2	4	8
x	2^{-3}	2^{-2}	2^{-1}	2^0	2^1	2^2	2^3
$\log_2 x$	-3	-2	-1	0	1	2	3
x	0.05	0.14	0.37	1	2.7	7.4	20.1
x	e^{-3}	e^{-2}	e^{-1}	e^0	e^1	e^2	e^3
$\log_e x$	-3	-2	-1	0	1	2	3

The logarithmic function for each of the three bases is plotted on the graphs.

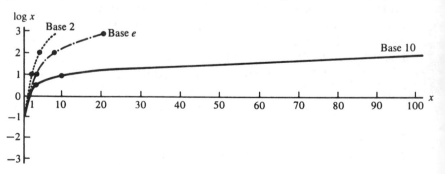

THE SIMPLE FUNCTIONS OF APPLIED MATH

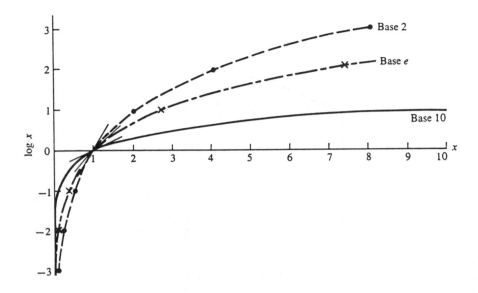

On the expanded graph the slope of each curve has been drawn at the point $x = 1$. As is the case with the *exponential* to the base e, there is something special about the slope (or first derivative) of the *log* to the base e. For this base the slope at $x = 1$ has the value of 1.

On the double graph below, $\log_e x$ is plotted with its slope, or first derivative values, directly below. At every point of $\log_e x$ the slope is equal to $1/x$. Hence, at $x = 1$ the slope is equal to 1.

$$\frac{d \log_e x}{dx} = \frac{1}{x} \tag{4-56}$$

The slope of $\log_2 x$ is greater than that of $\log_e x$, and the slope of $\log_{10} x$ is smaller. In general,

$$\frac{d \log_a x}{dx} = \frac{1}{x} \log_a e \tag{4-57}$$

For our two other common cases,

$$\frac{d \log_{10} x}{dx} = \frac{0.434}{x}$$

$$\frac{d \log_2 x}{dx} = \frac{1.44}{x}$$

What is "natural" about the natural log to the base e is that its first derivative is *equal* to $1/x$ and not just proportional to $1/x$. The standard nomenclature is that $\underline{\log_e x = \ln x}$.

4.4 THE LOGARITHMIC FUNCTION: $y = \log_b x$

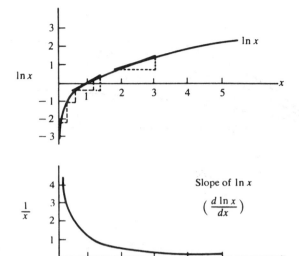

Note that every power function, x^a, is a derivative of some other power function, $[1/(a + 1)]x^{a+1}$, except for $a = -1$.

$$\frac{dx^2}{dx} = 2x^1 \qquad \frac{dx^1}{dx} = x^0 = 1 \qquad \frac{dx^{1/2}}{dx} = \tfrac{1}{2}x^{-1/2} \qquad \frac{d \ln x}{dx} = x^{-1}$$

$$\frac{dx^{-1}}{dx} = -x^{-2}$$

Transforming bases of logs

If $y = \log_a x$, then $x = a^y$.

Take the log of x to a different base, b,

$$\log_b x = \log_b (a^y) = y \log_b a = (\log_a x) \cdot (\log_b a) \tag{4–58}$$

For instance,

$$\log_e x = \log_{10} x \cdot \log_e 10 = 2.30 \log_{10} x$$
$$\log_{10} x = \log_e x \cdot \log_{10} e = 0.434 \log_e x$$

Another transformation of interest follows from letting $x = b$:

$$\log_b x = \log_a x \cdot \log_b a$$
$$\log_b b = \log_a b \cdot \log_b a$$

Since $\log_b b = 1$, $\log_a b = \dfrac{1}{\log_b a}$ \qquad (4–59)

C. The graphical relationship between the exponential and the log. Since the log of x to the base a is the exponent when x equals a raised to that

       THE SIMPLE FUNCTIONS OF APPLIED MATH

exponent, there is a reciprocal relationship between the log and the exponential. If $x = a^y$, then

$$y = \log_a x \tag{4–60}$$

These are two expressions for the *same* function. If we plot either one, we get the typical log curve shown on the graph below. If we interchange x and y, we have a different expression: $y = a^x$, and $x = \log_a y$. Again, both of these are expressions for the *same* function, which is plotted as the upper curve in the graph. The two curves are symmetrical about the 45° line. One is a mirror reflection of the other through that line, corresponding to an interchange of coordinates.

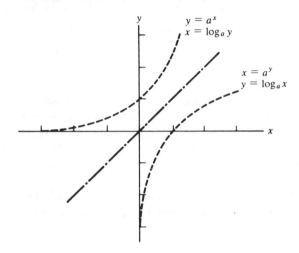

D. Uses of the logarithm

1. *Map of the universe:* We started the section on logarithms by defining them as the exponents in powers of 10, representing an array of numbers steadily increasing by factors of 10. For each increase of the number by a factor of 10 (a geometric progression), the exponent, or log, increased by 1 (an arithmetic progression). A prime use of such a functional relationship is the convenient representation of quantities that extend over a large range. For an example that is hard to beat, we present below a map of the sizes of objects in the universe, including the universe. The scale is logarithmic, necessarily, with the unit size being the meter. The linear dimension of each object portrayed is given by length $= (1 \text{ m}) \times 10^x$. Conversely, $x = \log_{10}$ (length in meters).

   Note particularly the position of man in this scheme of things and the small range occupied by living things. The scale extends no smaller than -15 because that is the current limit of meaningful measurement. At the other extreme there is indeed a limit to our universe. Since the velocity of objects away from us is proportional to their distance from us, there must exist a point beyond which objects are traveling at the speed of light. Regardless of what happens as that limit is approached, the fact remains that the light from those far-fleeing objects will be red shifted so far into the background noise that information from them will be irretrievably lost to us.

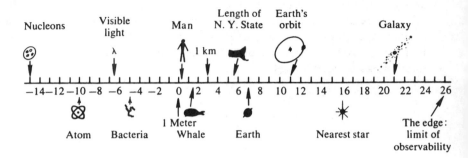

2. The slide rule: Although we derided the idea of using logarithms for computational purposes, we do indeed make use of logs every time we use a slide rule. A linear scale that can slide along another linear scale is an adding and subtracting machine. A log scale that can slide along another log scale is a multiplying and dividing machine, since adding logs is equivalent to multiplying numbers.

In the diagram we show the C and D scales of a standard slide rule. Attached to each of these log scales is a linear scale for comparison.

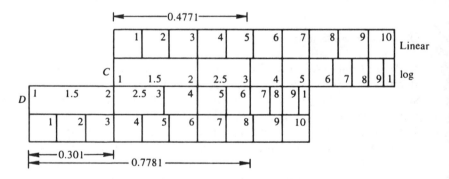

Note first that the linear scales do perform addition. The top scale starts at 3 of the bottom scale. Four places farther on, the top scale is directly over 7 on the bottom scale.

The C and D scales are proportional to logarithms. For instance, the log of 2 (to the base 10) is 0.3010. To multiply 2 and 3, add their logarithms and then take the antilog of that sum.

$$\log(2 \times 3) = \log 2 + \log 3 = 0.3010 + 0.4771 = 0.7781 = \log 6$$

On the slide rule this process is done without looking up the logs. By placing the 1 of the C scale over the 2 of the D scale the addition of logs begins with $\log 2 = 0.3010$. By adding the length on the C scale corresponding to 3, we have added 0.4771 linear units. The sum, 0.7781 linear units, corresponds to 6 on the D scale.

3. _The measure of acidity—pH:_ Pure water is a good electrical insulator because it has very few free ions. The dissociation of H_2O that does occur is into hydroxide ions (OH^-) and hydronium ions $(H_3O^+$ or larger clumps of H_2O attached to an extra H^+). The concentration of these is usually described by chemists in terms of molarity, in this case the number of moles of ions per liter of solution. A mole is best thought of as a large number: 1 mole $= 6 \times 10^{23}$. In pure water there are 10^{-7} mole of H_3O^+ ions per liter, or about 1 part in 5×10^8. (10^{-7} mole is equal to 6×10^{16}. In 1 liter of water there are 55.5 moles of H_2O molecules, or 3.3×10^{25} molecules.) There is an equal concentration of OH^- ions.

It might be thought that if more hydronium ions were poured into the water, in the form of HCl for instance, the small quantity of OH^- ions would all be neutralized, leaving no free hydroxide ions. Dissociation and recombination are always taking place, however, leading to a dynamic equilibrium. As the number of H_3O^+ increases by a factor of 10, the number of OH^- decreases by a factor of 10, but does not go to zero. The product of the two concentrations remains constant:

$$K_{\text{water}} = [OH^-][H_3O^+] = 1 \times 10^{-14} \qquad \text{at room temperature} \qquad (4\text{-}61)$$

Since the ion concentrations vary over such a large range, the natural mathematical description is in terms of logarithms. The <u>pH</u> (potency of hydrogen) of a water solution is defined as the negative log of the molar concentration of hydronium ions. For pure water the pH is $+7$, since there are 10^{-7} mole of H_3O^+ per liter. The molar concentration of H_3O^+ in soda pop is 10^{-3}; hence the pH is $+3$, and the molar concentration of OH^- is 10^{-11}. In 0.1 molar (M) sodium bicarbonate the pH is 8.4, therefore the molar concentration of OH^- ions is $10^{-5.6}$. In this case the hydroxide ions dominate and the solution is basic, not acidic. The range of pH values for some common substances is

1.0 M HCl	0.1	rainwater	6.2
0.1 M HCl	1.0	milk	6.5
gastric fluid	2.0	seawater	8.5
lemon juice	2.3	milk of magnesia	10.5
vinegar	2.8	0.1 M NaOH	13.0
tomatoes	4.2	1.0 M NaOH	14.0

4. _The senses—decibels:_ Most measuring instruments are sensitive only over a very narrow range. Voltmeters rarely have more than 100 divisions on their scale, and the fractional precision, of course, is inversely proportional to the magnitude of the reading. Even the standard meter stick has only 1000 scale divisions. As we move from one range to another of any variable that we are measuring, we must change instruments, or at least the scale of the instrument.

Fortunately for us, the senses of sight, hearing, feeling, etc., are exceptions to this rule. Although our precision is usually not great, still we must be able to sense forces from 10^{-3} N to 10^{+3} N, from the weight of an insect to the weight of a man. We can see to some extent in starlight, when the illuminance is only 10^{-3} lumens/m^2, and can also operate in bright sunlight at 10^5 lumens/m^2. Our sense of hearing covers a range of 10^{12}. Furthermore, all these senses respond in a roughly logarithmic fashion. Not only do they cover a large range of magnitude, but their fractional precision remains about the same throughout the range. To illustrate this characteristic of logarithmic response, consider (or measure!) your sensitivity to differences in weights. Arrange the experiment so that weights are held by a thread in each hand and see whether you can detect a difference in weight if 1 g is held in one hand and 0.9 g in the other. When you are convinced that you can tell a difference of 0.1 g (0.001 N), or 0.2 g, see whether or not you can detect the same difference if one hand holds 1 kg and the other holds 999.9 or 999.8 g. Of course, you cannot. What you will find is that your *fractional* sensitivity remains about the same. If you can detect a difference of 1 part in 10 in one part of the range, that is approximately your sensitivity in any other part. The math model of this characteristic is that for some variable x you perceive a response y, where $y = k \log x$. A change in the variable produces a change in the response as follows:

$$\Delta y = K \frac{1}{x} \Delta x \qquad \left[\text{since } \frac{d(\log x)}{dx} \propto \frac{1}{x} \right]$$

Equal increments of response, Δy, correspond to equal *fractional* increments, $\Delta x/x$, of the variable.

Corresponding roughly to human perception, the intensity of sound is measured on a logarithmic scale. In terms of intensity I, measured in watts (W)/cm^2, the intensity level is defined to be

$$(\text{decibels}) \ \mathrm{dB} = 10 \log_{10} \frac{I}{I_0} \tag{4-62}$$

On such a scale there must be a unit reference intensity. I_0 is usually chosen to be 10^{-16} W/cm^2, which is around the threshold of human hearing at 300 hertz (Hz) (cycles/second). The unit of intensity level is the decibel. Originally, intensity levels were defined in terms of $\log_{10} I/I_0$, and the unit was called the bel in honor of Alexander Graham Bell. The decibel is more convenient because the human ear can detect a difference in intensity level of about 1 dB. Note that if the actual energy intensity doubles, the ear hears an increase of 3 dB.

THE SIMPLE FUNCTIONS OF APPLIED MATH

The intensity levels of various sounds are given below. They are only approximate because of the qualitative method of describing them and because the sense of hearing is strongly dependent on the frequency of the sound.

Sound	Level (dB)	$I(W/cm^2)$
Threshold	0	10^{-16}
Whisper	20	10^{-14}
Quiet office	40	10^{-12}
Conversational speech	60	10^{-10}
Heavy traffic	70	10^{-9}
Passing subway	90	10^{-7}
Hammering machinery	110	10^{-5}
Pain threshold	120	10^{-4}

If the absolute intensity is measured in terms of the *pressure* of the sound, then the sound level in decibels is equal to $20 \log p/p_0$, since the energy flow is proportional to the square of the pressure amplitude of the wave. There is a similar situation in electrical engineering measurements of electromagnetic energy flow. The decibel amplification is equal to $10 \log E/E_0$ or $20 \log V/V_0$, where E is the measured energy and V is the voltage.

5. Stellar magnitudes: About 150 B.C. a Greek named Hipparchus, working at Rhodes and Alexandria, classified stars according to their brightness. He divided them into six groups or magnitudes. As newer methods of measuring intensity have been developed during the last few centuries, the scales have always been adjusted to make them compatible with previous scales, which in turn trace back to Hipparchus. Consequently, the modern logarithmic scale of stellar intensities has five magnitudes for a change in intensity of 100. This leads to a log scale with a base of $\sqrt[5]{100} = 2.51$. Furthermore, so as not to offend Hipparchus or all the catalogues based on the work based on his work, the scale is reversed so that large magnitudes correspond to small brightness. The reference level is also historical and arbitrary, but on that scale the dimmest star that we can see with the unaided eye has magnitude $+6$, the brightest star, Sirius, has magnitude -1.5, the sun has magnitude -26.9, and the full moon is -12.7. The faintest object detected with the 200-in. telescope has apparent magnitude of $+23$. The formula for the difference between magnitudes is

$$b - a = \log_{2.51} \frac{I_a}{I_b} = 2.5 \log_{10} \frac{I_a}{I_b} \tag{4-63}$$

If star a is 10 times as bright as star b, then star b is 2.5 magnitudes greater than star a. Note that a factor of 100 in intensity yields a factor of 5 in magnitude.

The actual ranking of stars according to brightness is very complicated. The color of the starlight and the frequency detection efficiency of the receiver must be taken into account. Furthermore, the magnitudes given above are "apparent" magnitudes as seen from the earth. More important for astrophysics is the intrinsic magnitude of a star, the magnitude that it would have if the star were at some unit distance from the earth. The apparent magnitude is converted to the absolute magnitude by calculating the brightness that the star would have if it were at 10 parsecs (about 33 light-years) from the earth. To do this calculation, of course, the distance to the star must be known. The absolute visual magnitude (matching the spectral sensitivity of the human eye) of Sirius is $+1.4$, that of the sun is $+4.7$, and the moon would be $+32$. One of the brightest stars known is Rigel in Orion with an absolute visual magnitude of -7.0.

6. *Electric potential from a line source:* From geometrical arguments about the distribution of lines of force (see p. 35), the electric field produced by a long cylinder or line of electric charge can be deduced to be $E \propto q/r$, where q is the charge per unit length and r is the radial distance to the center of the line source. The relationship between electric field, E, and electric potential, V, is given by $E_r = -(dV/dr)$. Since we know that $E \propto 1/r$ for a line source, the potential can be calculated by asking what function of r has a first derivative proportional to $-1/r$? The function, of course, if $-\log r$. As usual with potentials, there has to be a reference point, or else only potential differences can be considered. For a point or spherical source the zero potential is usually chosen to be at infinity. In the case of $V \propto -\log r = \log 1/r$, the potential does not go to zero as r goes to infinity. It is just as well, since the math would have to be the model for an infinite line source. Only potential differences are meaningful in this case:

$$V_a - V_b \propto (-\log r_a + \log r_b) = \log \frac{r_b}{r_a} \qquad (4\text{-}64)$$

$V_a - V_b$ is positive if the source charge is positive.

SAMPLES

1. Using the technique of taking roots, fill in the ln values for the interval between 1 and e.

2. The $\log_{10}$ of $34 = \log_{10}(3.4 \times 10^1) = 1 + \log_{10} 3.4$. Find the equivalent method of writing the ln of 34, so that one needs only the ln values between 1 and e.

3. Use the log tables to find the value of $(8432 \times 0.00631)/(180.2 \times 0.842)$. (3.51×10^{-1})

4. Carefully plot $\log_{10} x$ and $\ln x$ from $x = 0.1$ to $x = 3$. Draw and measure the slope of each at $x = 1$.

5. Transform $\log_2 x$ to $\ln x$ and to $\log_{10} x$.

6. With an input signal of 0.1 volt (V), an audio amplifier produces $10\,V$ output up to 10,000 Hz. From that point on the response falls almost linearly to 0 at 12,000 Hz. At what frequency has the response fallen off by 3 dB? (10,600 Hz.)

7. The apparent magnitude of our sun is about -26.9. Its absolute magnitude is $+4.7$. Its actual distance from us is about 8 light-minutes. The standard distance for determination of absolute magnitude is about 33 light-years. Check the consistency of all this. (Remember the inverse-square law.)

4.5 THE FACTORIAL: $y = n!$ The factorial appears in many expressions concerned with probability or distributions.

$$n! = n(n - 1)(n - 2) \cdots 3 \cdot 2 \cdot 1 \qquad \text{where } n \text{ is a positive integer} \qquad (4\text{--}65)$$

The factorial will outrun any of the other functions that we have described. For instance, $a^n/n! \longrightarrow 0$ as $n \longrightarrow \infty$. Because $n!$ grows so rapidly, we graph it on a semilog plot:

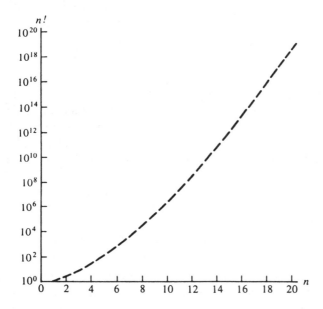

Notice that the function is growing faster than an exponential, which would produce a straight line on a semilog plot. Notice also how enormous the values are even for relatively small n: $10! = 3,628,800$ and $100! = 9.33 \times 10^{157}$.

There is a useful approximation to the factorial, called Stirling's formula.

$$n! \simeq n^n e^{-n} \sqrt{2\pi n} = \left(\frac{n}{e}\right)^n \sqrt{2\pi n} \qquad (4\text{--}66)$$

This is an asymptotic approximation. The absolute error increases as n increases, but the percentage error decreases. It is surprisingly good, even for very small n. For example, for $n = 1$, Stirling's formula gives $1! = (1/e)\sqrt{2\pi} = 0.924$. The error is less than 8 per cent. For $n = 10$, Stirling's formula gives $10! = (10/e)^{10}\sqrt{20\pi} = 3,598,696$. The error in this case is 0.8 per cent. For $n = 100$, the error is 0.08 per cent.

SAMPLES

1. Find the percentage error in using Stirling's formula when $n = 5$. (1.3 per cent.)

2. How many combinations of different tests of 10 questions each can be made out of 30 test questions? The formula is $_nC_r = n!/[r!(n - r)!]$. (See Section 4.6 on the binomial distribution.) This formula counts (1, 2, 3, 4, 5, 6, 7, 8, 9, 10) as a different test from (11, 2, 3, 4, 5, 6, 7, 8, 9, 10), but counts (1, 2, 3, 4, 5, 6, 7, 8, 9, 10) as the same test as (2, 1, 3, 4, 5, 6, 7, 8, 9, 10). Use Stirling's approximation to save time. Reduce the numerator and denominator terms by cancellation.

$$\left(\frac{30}{e}\right)^{30} = \left(\frac{3}{e}\right)^{30}(10)^{30} \quad \text{and} \quad \left(\frac{10}{e}\right)^{30} = \left(\frac{10}{e}\right)^{10}\left(\frac{10}{e}\right)^{20}$$

Remember that $a^x = b^{x \log_b a}$ [formula (4–45)]. (3.1×10^7.)

4.6 THE BINOMIAL DISTRIBUTION: $P_r = {}_nC_r p^r(1 - p)^{n-r}$ (4–67)

The most probable number of heads expected in 20 tosses of a coin is 10, but there is a finite chance of getting any other number of heads from 0 to 20. In many trials of 20 tosses each you would expect to have a distribution of numbers centered around 10. The binomial distribution describes that expectation. The probability of getting r heads in n throws is p_r, a number always less than 1. The probability of "success" in a single event is p. In the case of flipping a coin, the chance of getting a head on a single toss is $p = \frac{1}{2}$. The chance of getting any particular number from 1 to 6 on a cast of a single die is $p = \frac{1}{6}$. The probability of "failure" in a single event is $1 - p$.

A. The symbol $_nC_r$ stands for the number of *combinations* of n things taken r at a time *when no account is taken of the sequence of members within a group.* Suppose that we count the number of ways in which 10 objects can form doublets, *if we count a doublet such as (8, 5) as being different from (5, 8)*. There are 10 choices for the first member of each doublet and after that there are 9 choices for the second member. The total number of choices, 90, can be written as $10 \cdot 9$ or as $10!/8!$. That includes the doublets (8, 5) *and* (5, 8). In general, with n objects assembled into groups of r each, the number of choices available is $n!/(n - r)!$, *if* each permutation of the members within each group is counted as a separate group. But now, within each group

THE SIMPLE FUNCTIONS OF APPLIED MATH

with r members there are $r!$ such permutations. [For $r = 3$, $r! = 6$. $(1, 2, 3)$, $(1, 3, 2)$, $(2, 1, 3)$, $(2, 3, 1)$, $(3, 1, 2)$, $(3, 2, 1)$.] Therefore, the number of combinations of n things taken r at a time *when no account is taken of the sequence of members within a group is*

$$_nC_r = \frac{n!}{r!(n-r)!} \tag{4-68}$$

In terms of our original example, $_{10}C_2 = 10!/(2!\ 8!) = 45$, because now we call $(8, 5)$ the same as $(5,8)$.

The distribution is called binomial because p_r is also a term in the expansion of $(q + p)^n$, where $q = 1 - p$.

$$(q + p)^n = q^n + nq^{n-1}p + \frac{n(n-1)}{2}q^{n-2}p^2 + \cdots$$
$$+ {}_nC_r p^r(1 - p)^{n-r} + \cdots + p^n$$

The arguments leading to the probability distribution are as follows: If on a single draw or cast the probability of success is p ($\frac{1}{2}$ for the coin, $\frac{1}{6}$ for the die), then the probability of having successes for two throws in a row is p^2. The probability of having successes for the first two throws out of three, and then a failure, is $p^2(1 - p)$. However, the probability of having exactly two successes out of three throws is equal to the sum of the probabilities of all sequences that can yield two successes. There are three such sequences, each with the same independent probability: (success, success, failure); (s, f, s); (f, s, s). The total probability of obtaining two successes out of three throws is thus $3p^2(1 - p)$. In general, for n trials each specific sequence that can yield r successes will have a probability of $p^r(1 - p)^{n-r}$. Out of n trials, the number of sequences with r successes is the number of ways n things can form groups of r, which is $_nC_r$. Therefore, the total probability that there will be exactly r successes in n independent trials is

$$_nC_r p^r(1 - p)^{n-r} = \frac{n!}{r!(n-r)!}p^r(1 - p)^{n-r}$$

The most likely number of successes is the greatest integer less than $p(n + 1)$. If $p(n + 1)$ is an integer, there are two most likely numbers, $p(n + 1)$ and $p(n + 1) - 1$. Approximately, the maximum of the distribution curve is at np.

The probability of obtaining the most likely number is given by setting $np = r$: $p_{np} = {}_nC_{np}p^{np}(1 - p)^{n-np}$. Using Stirling's formula for $n!$, $np!$, and $(n - np)!$, we find that the probability of getting the most likely number is approximately

$$p_{np} \simeq \frac{1}{\sqrt{2\pi np(1 - p)}} \tag{4-69}$$

B. Here are distribution curves for two different values of p, each for three values of n. Notice that as n increases, the probability of any particular num-

4.6 THE BINOMIAL DISTRIBUTION: $P_r = {}_nC_r p^r(1 - p)^{n-r}$

ber r decreases. In particular, the chance of getting the most likely number decreases. As n increases, the distribution around the most likely number shrinks in a relative sense but not in an absolute sense. For larger n there is greater chance of missing the most likely number by any absolute amount.

For $p = 0.2$ the most likely number is 1 for $n = 5$, 2 for $n = 10$, and 5 for $n = 25$. Notice the way the distributions are skewed near the origin. Relative symmetry increases as n increases and takes np farther away from the origin.

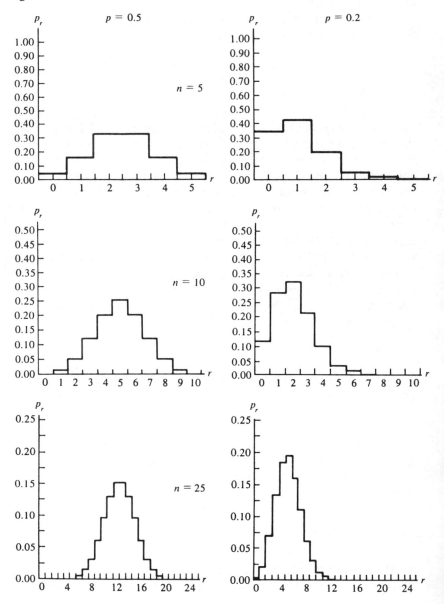

THE SIMPLE FUNCTIONS OF APPLIED MATH

Two approximations to the binomial distribution are particularly important. The first is Poisson's formula for the occurrence of small numbers; the second is the normal curve for large numbers, sometimes known as Laplace's approximation or, with a slight change of variable, the Gaussian error function.

C. Poisson Distribution: Replace $n!$ and $(n - r)!$ in the binomial distribution with the approximate values from Stirling's formula. Make the approximation that $r \ll n$, that $n - r + \frac{1}{2} \simeq n$, that $[1 - (r/n)]^n \simeq e^{-r}$, and that $(1 - p)^n \simeq e^{-np}$. The result is

$$p_r = \frac{(np)^r}{r!} e^{-np} \qquad (4\text{-}70)$$

Poisson's formula applies to rare events. The approximation gets better as p gets smaller and n gets larger, the product np remaining the same.

Here is a comparison of Poisson's approximation and the actual values of the binomial for $np = 2$, first with $p = 0.2$ and then with $p = 0.1$. The second is clearly a better approximation.

	$n = 10, p = 0.2$			$n = 20, p = 0.1$	
r	p_r(binomial)	p_r(Poisson)	r	p_r(binomial)	p_r(Poisson)
0	0.107	0.135	0	0.121	0.135
1	0.268	0.270	1	0.270	0.270
2	0.303	0.270	2	0.285	0.270
3	0.202	0.180	3	0.190	0.180
4	0.088	0.090	4	0.090	0.090
5	0.026	0.036	5	0.032	0.036
6	0.005	0.010	6	0.009	0.012
7	0.001	0.003	7	0.002	0.003
8	0.000	0.001	8	0.000	0.000
9	—	—	9	—	—
10	—	—	10	—	—

As an example of the use of Poisson's formula, suppose that a radioactive sample gives an average of 3 counts per second. What is the probability of getting r counts per second? Since the expected number is $np = 3$, $p_r = (3^r/r!)e^{-3}$.

r	p_r	r	p_r
0	0.050	6	0.051
1	0.149	7	0.022
2	0.224	8	0.008
3	0.224	9	0.003
4	0.169	10	0.001
5	0.101		

4.6 THE BINOMIAL DISTRIBUTION: $P_r = {}_nC_r p^r (1 - p)^{n-r}$

Notice that there is only a 22 per cent chance of getting 3 counts during any 1 second, a 5 per cent chance of getting no counts; and a 1 per cent chance of getting 8.

D. The normal distribution, or Laplace's approximation, or the Gaussian error curve: If n, $n - r$, and r are all large enough to justify the use of Stirling's formula, the binomial distribution becomes

$$p_r = \frac{1}{\sqrt{2\pi npq}} e^{-x^2/2npq} \qquad \text{where } q = 1 - p, \text{ and } x = r - np \qquad (4\text{--}71)$$

The substitution of x for $r - np$ emphasizes that the formula is concerned with deviations of r from the central, or expected value, np. This curve is the same as the Gaussian error curve, described on p. 87, if we set $\sigma = \sqrt{npq}$. Then, $p_r = (1/\sigma\sqrt{2\pi})e^{-x^2/2\sigma^2}$. When $r = np$, $x = 0$, and $p_r = 1/\sigma\sqrt{2\pi} = 1/\sqrt{2\pi npq}$, as we showed earlier. The integral of p_r from minus infinity to plus infinity is just equal to 1, as it must be if it represents the total probability of an event taking place for some r.

Notice that the standard deviation is proportional to $\sqrt{n}$. The standard deviation measures the "width" of the curve. As n increases, this width increases, but the relative width, σ/n, decreases.

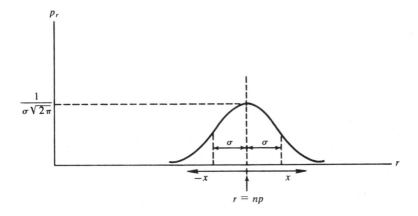

SAMPLES

1. What is the most likely number of combinations adding to five to get in 72 casts of two dice? (8.) What is the probability of getting 4 of them? (5.8×10^{-2}.) 2 of them? (1.1×10^{-2}.) 8 of them? (15×10^{-2}.)

2. A particular radioactive sample produces an average of 100 counts/min. What is the probability that, during a particular count of 1 min., exactly 100 counts will be obtained? (4.0×10^{-2}.) What is the probability that a count of exactly 90 will be obtained? (2.4×10^{-2}.) What is the probability that a count between 90 and 110 will be obtained? (68×10^{-2}.)

THE SIMPLE FUNCTIONS OF APPLIED MATH

4.7 HYPERBOLIC FUNCTIONS—sinh *u*, cosh *u*, tanh *u*

A. Definition and Relationships. The hyperbolic functions consist of the sine, cosine, tangent, etc., derived from the unit hyperbola, $x^2 - y^2 = 1$. As their names imply, they are closely analogous to the ordinary sinusoidal or circular functions that can be derived from the unit circle, $x^2 + y^2 = 1$. Every point on the unit circle can be given in terms of a parameter, θ: $x = \cos\theta$, $y = \sin\theta$. Similarly, any point on the unit hyperbola can be expressed as

$$x = \cosh u, \quad y = \sinh u. \tag{4-72}$$

(The names are pronounced "cosh" and "sinch.") The other hyperbolic functions are related in the same way as the corresponding sinusoidal functions:

$$\tanh u = \frac{\sinh u}{\cosh u}, \quad \operatorname{sech} u = \frac{1}{\cosh u}, \quad \operatorname{cosech} u = \frac{1}{\sinh u},$$

$$\coth u = \frac{1}{\tanh u} \tag{4-73}$$

The graphs of sinh *u*, cosh *u*, and tanh *u* are given in the appendix

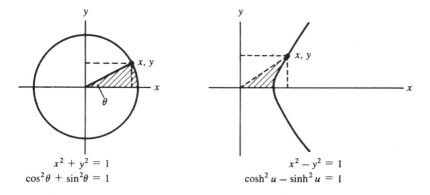

$$x^2 + y^2 = 1 \qquad\qquad x^2 - y^2 = 1$$
$$\cos^2\theta + \sin^2\theta = 1 \qquad \cosh^2 u - \sinh^2 u = 1$$

The parameter, *u*, is not the angle between the *x* axis and the radius arm to the point on the hyperbola, but there is a correspondence between the significance of θ for the circle and *u* for the hyperbola. The area swept out by the radius of the circle in going from $(0, 0)$ to (x, y) is $\frac{1}{2}r^2\theta$, which for the unit circle has the value $\frac{1}{2}\theta$. Similarly, the area swept out by the radius arm to the hyperbola is equal to.

$$A = \tfrac{1}{2}xy - \int_1^x \sqrt{x^2 - 1}\, dx$$
$$= \tfrac{1}{2}xy - [\tfrac{1}{2}x\sqrt{x^2 - 1} - \tfrac{1}{2}\ln(x + \sqrt{x^2 - 1})]_1^x \tag{4-74}$$
$$= \tfrac{1}{2}xy - \tfrac{1}{2}x\sqrt{x^2 - 1} + \tfrac{1}{2}\ln(x + \sqrt{x^2 - 1})$$

Since $y = \sqrt{x^2 - 1}$,

$$A = \tfrac{1}{2} \ln(x + y)$$

or

$$x + y = e^{2A}$$

Since $x^2 - y^2 = 1$, it follows that

$$x - y = e^{-2A}$$

Add and subtract these two expressions to obtain

$$x = \tfrac{1}{2}(e^{2A} + e^{-2A})$$
$$y = \tfrac{1}{2}(e^{2A} - e^{-2A})$$

If we define the hyperbolic parameter, u, to have a value equal to $2A$, in analogy with the circular relationships, we can define the hyperbolic sine and cosine in terms of these exponential functions ($u = 2A$ for the *unit* hyperbola: $u = 2A/a^2$ when $x^2 - y^2 = a^2$):

$$x = \tfrac{1}{2}(e^u + e^{-u}) \equiv \cosh u$$
$$y = \tfrac{1}{2}(e^u - e^{-u}) \equiv \sinh u \tag{4-75}$$

Note the analogy with the circular functions:

$$x = \tfrac{1}{2}(e^{i\theta} + e^{-i\theta}) \equiv \cos \theta$$
$$y = \frac{1}{2i}(e^{i\theta} - e^{-i\theta}) \equiv \sin \theta \tag{4-76}$$

Also note that

$$\cosh^2 u - \sinh^2 u = \tfrac{1}{4}(e^{2u} + e^{-2u} + 2e^0) - \tfrac{1}{4}(e^{2u} + e^{-2u} - 2e^0)$$
$$= 1 \tag{4-77}$$

and

$$\cosh i\theta = \tfrac{1}{2}(e^{i\theta} + e^{-i\theta}) = \cos \theta$$
$$\sinh i\theta = \tfrac{1}{2}(e^{i\theta} - e^{-i\theta}) = i \sin \theta \tag{4-78}$$

The cosh and sinh of the sum of two parameters are only slightly (but very importantly) different from those of the trigonometric counterparts.

$$\cosh(u \pm v) = \cosh u \cosh v \pm \sinh u \sinh v$$
$$\sinh(u \pm v) = \sinh u \cosh v \pm \cosh u \sinh v \tag{4-79}$$

These can be figured out from the formulas given for cosh u and sinh u, but if you remember their values at 0 (cosh $0 = 1$, sinh $0 = 0$) and keep in mind that they increase with increasing angle, you can figure out the appropriate signs.

As you can demonstrate with the exponential formulas,

$$\frac{d}{du} \cosh u = \sinh u \qquad \text{and} \qquad \frac{d}{du} \sinh u = \cosh u \tag{4-80}$$

The series expansions for $\cosh u$ and $\sinh u$ are

$$\cosh u = 1 + \frac{u^2}{2!} + \frac{u^4}{4!} + \frac{u^6}{6!} + \cdots$$

$$\sinh u = u + \frac{u^3}{3!} + \frac{u^5}{5!} + \frac{u^7}{7!} + \cdots \qquad -\infty < u < \infty \qquad (4\text{-}81)$$

See Appendix 9 for graphs of the hyperbolic sine, cosine, and tangent.

B. An Application of Hyperbolic Functions. Besides being of mathematical interest, the hyperbolic functions figure prominently in Einstein's special theory of relativity. The transformation equations relating x and t to x' and t' are

$$x' = \frac{x - vt}{\sqrt{1 - (v^2/c^2)}} \qquad t' = \frac{t - (xv/c^2)}{\sqrt{1 - (v^2/c^2)}} \qquad (4\text{-}82)$$

The coordinates x' and t' are for a reference frame moving with velocity v with respect to the frame specifying x and t. There is an invariant interval that has the same value for all frames moving at constant velocity (in the x direction) with respect to each other:

$$x'^2 - c^2t'^2 = x^2 - c^2t^2 \qquad (4\text{-}83)$$

For a particular value of this interval, the relationship between x and t in any frame is hyperbolic. For instance, suppose that we choose a moment in the (x', t') frame when $t' = 0$ and $x' = 1$. Then in the (x, t) frame the relationship between x and t must be such that

$$1 = x^2 - c^2t^2$$

The values for x and t lie along a hyperbola where $x = \cosh u$ and $ct = \sinh u$.

These relationships can best be seen in a graphical form that was first used by Minkowski in 1908. A stationary point on the x axis traces out a

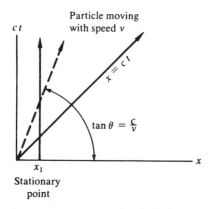

vertical "world line," with a velocity c, corresponding to the passage of time, t. If a particle has velocity it also moves along the x axis at speed v while

4.7 HYPERBOLIC FUNCTIONS—$\sinh u$, $\cosh u$, $\tanh u$

being swept along the ct axis at speed c. The slope of its world line is c/v $= 1/\beta$. The world line of a light beam has a slope of 1 since $\tan \theta = c/c = 1$.

The primed reference frame can be shown on the same diagram. The ct' and x' axes must be symmetrical about the world line of the light beam, since in all reference frames the velocity of light must be the same. This is assured if $x_1/ct_1 = 1$ and $x'_1/ct'_1 = 1$.

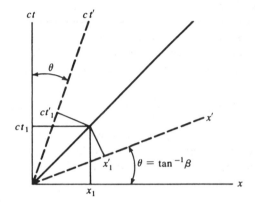

The "metrics" of the two sets of axes are not the same. A length of x on the graph does not correspond in value to the same length x' on the graph. Instead, they are related by the hyperbolic requirements of the invariant interval, $x^2 - c^2t^2$. The graphical result is shown in the diagram. The

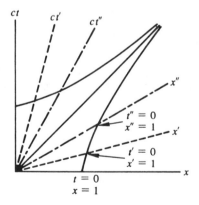

upper hyperbola is drawn for the case where ct' is chosen to be equal to 1 and $x' = 0$. Then $-1 = x^2 - c^2t^2$ or $1 = c^2t^2 - x^2$. The faster the other reference frames travel with respect to us, the more their distance and time scales appear to be warped. The choice of unity for x' and t' was arbitrary; to continue the metric calibration we could draw hyperbolas for $x' = 2$, $t' = 2$, etc. There is no need to, however, because the metric is necessarily linear in any inertial frame.

THE SIMPLE FUNCTIONS OF APPLIED MATH

To determine the amount of space warping, draw a vertical world line from the x axis through the point, $t'' = 0, x'' = 1$. This world line traces one end point of a measuring rod extending from 0 to about $1\frac{1}{4}$ units in the x frame. In the x'' frame, however, it has a length of only 1 unit. In other words, a rod 125 cm long in the x frame is only 100 cm long in this x'' frame. To see the warping in the time dimension, draw a vertical world line from the x axis through the point, $ct'' = 1, x'' = 0$. If this represents a time of one hour in the t'' frame, it represents about $1\frac{1}{4}$ hours in the t system. This warping is the cause of the lengthened lifetime of the traveling twin in the famous twin "paradox." While the stay-at-home twin lives an hour and a quarter, his traveling twin in this particular example would have lived for only one hour.

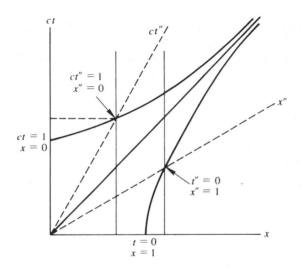

The relationships among hyperbolic functions and the use of the Minkowski diagram provide an easy derivation of the relativistic formula for addition of velocities. Recall that a point on the hyperbola can be given in terms of a parameter u. [For the Minkowski diagram, $u = \ln\sqrt{(1 + \beta)/(1 - \beta)}$, but we do not need this value for the next step.]

$$x = \cosh u \qquad ct = \sinh u$$

$$v = \frac{dx}{dt} = \sinh u \frac{du}{dt} \qquad \frac{dt}{du} = \frac{1}{c} \cosh u$$

Therefore, the velocity of the moving frame is

$$v = c \tanh u \qquad\qquad (4\text{–}84)$$

From within the primed frame the velocity of the double primed frame is $v' = c \tanh w$. The velocity of the double primed frame as seen in the stationary frame is *not* $v + v'$, but is given by

4.7 HYPERBOLIC FUNCTIONS—sinh u, cosh u, tanh u

$$v'' = c \tanh(u + w)$$

$$v'' = c \tanh(u + w) = c\frac{\sinh(u + w)}{\cosh(u + w)} \tag{4-85}$$

$$= c\frac{\sinh u \cosh w + \cosh u \sinh w}{\cosh u \cosh w + \sinh u \sinh w}$$

Divide numerator and denominator by $\cosh u \cosh w$:

$$v'' = c\frac{\tanh u + \tanh w}{1 + \tanh u \tanh w} \tag{4-86}$$

$$v'' = \frac{v + v'}{1 + (vv'/c^2)}$$

This is the addition law for velocities.

See p. 258 for another application of the hyperbolic function.

SAMPLES

1. Use the exponential notation to demonstrate that $\cosh(u \pm v) = \cosh u \cosh v \pm \sinh u \sinh v$.

2. Use the exponential notation to demonstrate that $d/du \cosh u = \sinh u$ and $d/du \sinh u = \cosh u$.

3. Use the Minkowski diagram in the illustration to determine approximately the angle of the world line of the reference frame (and hence its velocity) that has a length contraction of a factor of 2 with respect to us. ($\theta \approx 40° = \tan^{-1} \beta$; therefore, $\beta \approx 0.34$.)

CHAPTER5 STATISTICS

5.1 DEFINITIONS AND PROPERTIES OF THE GAUSSIAN. It is well known that figures can lie and liars can figure. The method most commonly used is called statistics. Statistics is the mathematical treatment and analysis of large quantities of data. It provides us with ways of characterizing the distribution of such data with a few parameters, such as averages. It also establishes rules for drawing samples from large amounts of data. All these rules are based on certain laws of probability and on requirements about the random nature of obtaining particular values of the variables. When the prior conditions on the data are not satisfied, the magnificent mathematical paraphernalia can produce errant nonsense. Statistics is the main analytical tool of some of the social sciences, particularly those connected with research in education.

A. A complete body of data to be analyzed is called the *population*. This might be a large but finite list, such as the IQ scores for all students in the United States. It could also be an infinite body of data, such as the distribution of the sum of the dots registered on two dice for an infinite number of throws. For many purposes we must be satisfied with taking a sample of the population. Then we must know how likely it is that an average or some other characteristic of the sample is equal to that characteristic of the population.

The data, whether from the population or the sample, can be presented graphically as a distribution. The horizontal axis is used for the *random variable*. In the case of the IQ grades, the random variable is the grade itself, running from 0 to 200. In the case of the dice, the random variable is the sum of the dots, running from 2 to 12. The vertical axis is used for the frequency of occurrence. The frequency might be given in terms of the fraction of all possible events, or in terms of the actual number in some particular sample. If the random sample is quantized, the distribution graph is a histogram. If the random variable is continuous, then the distribution graph will contain a smooth curve. A distribution can also be described by a function, $f(x)$.

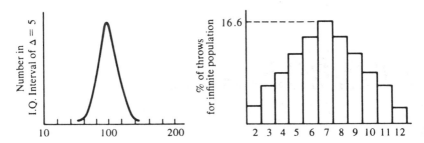

The distribution graph is thus a plot of $f(x)$ versus x.

Distributions can have any form from equal values for all values of the variable (the graph would be a horizontal line) to zero value for every region of the variable except one (the graph would be a thin vertical peak at that region). In a surprising number of situations, distribution graphs are bell shaped. There is a maximum value of occurrence for one region of the random variable and rapidly decreasing probability of finding larger or smaller values.

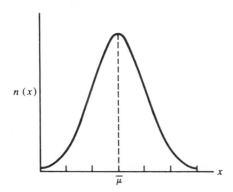

B. In many cases experimental data can be fitted to a very particular form of bell-shaped curve called the *Gaussian distribution*. The formula for the Gaussian distribution is

$$n(x) = \frac{N}{\sigma\sqrt{2\pi}}e^{-\frac{(\bar{a}-x)^2}{2\sigma^2}} \qquad (5\text{--}1)$$

N is the total number, and $n(x)$ is the value of the distribution function at x. (It is the number of events between x and $x + dx$.) The *arithmetic average*, or *mean*, of the distribution is $\bar{\mu}$, and σ is a measure of the width or spread of the curve. Note the qualitative behavior of the Gaussian function. It has its maximum value when $x = \bar{\mu}$. At that point $n(x) = N/\sigma\sqrt{2\pi}$ since $e^0 = 1$. The function is symmetric about $\bar{\mu}$ because $\bar{\mu} - x$ is squared. As x gets larger or smaller than $\bar{\mu}$, the value of the Gaussian function rapidly decreases so that it is nearly zero if the exponent is much larger than 3 or 4. The numerical value of the exponent depends on the ratio of the square of the departure from $\bar{\mu}$, $(\bar{\mu} - x)^2$, to the denominator, $2\sigma^2$. Evidently the quantity $2\sigma^2$ determines the sharpness of the falloff and so is a measure of the width of the curve. If σ is very large, it takes a large value of $\bar{\mu} - x$ to reduce the exponential. The quantity, σ, is called the *standard deviation*.

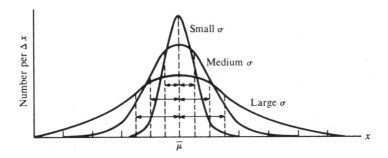

It is not obvious from looking at the formula, but necessarily the total area under the curve, $\int_{-\infty}^{\infty} n(x)dx$, must equal the total number, N. It is indeed the case that

$$\int_{-\infty}^{\infty} \frac{1}{\sigma\sqrt{2\pi}} e^{-(\bar{\mu} - x)^2/2\sigma^2} dx = 1.$$

Integrals of the Gaussian between other limits have special significance.

1. $\int_{\bar{\mu}-\sigma}^{\bar{\mu}+\sigma} \frac{N}{\sigma\sqrt{2\pi}} e^{-(\bar{\mu}-x)^2/2\sigma^2} dx = 0.68N$ (5–2)

   This gives the area (or number of events) under the curve between plus and minus *one* standard deviation around the mean. Note that about two thirds of the events occur within that region.

2. $\int_{\bar{\mu}-1.5\sigma}^{\bar{\mu}+1.5\sigma} \frac{N}{\sigma\sqrt{2\sigma}} e^{-(\bar{\mu}-x)^2/2\sigma^2} dx = 0.87N$ (5–3)

3. $\int_{\bar{\mu}-2\sigma}^{\bar{\mu}+2\sigma} \frac{N}{\sigma\sqrt{2\pi}} e^{-(\bar{\mu}-x)^2/2\sigma^2} dx = 0.95N$ (5–4)

4. $\int_{\bar{\mu}-3\sigma}^{\bar{\mu}+3\sigma} \frac{N}{\sigma\sqrt{2\pi}} e^{-(\bar{\mu}-x)^2/2\sigma^2} dx = 0.997N$ (5–5)

Note that only 0.3 per cent of the events occur outside a range of plus or minus three standard deviations from the mean.

Often the random variable is not continuous, such as with the distribution of dice throws. Even when a continuous range of variable does occur, we frequently group the data in bins. All the readings between 0 and 4.9 s go in one bin, between 5 and 9.9 in the next, etc. The definitions for the mean and standard deviation in terms of the raw data are given as follows. The symbol, $\sum_i$, means a summation of all the terms in the bins labeled $i = 1, 2, 3$, etc. In bin i there are n_i occurrences, each with the average value x_i.

1. *Arithmetic average or mean.*

$$\bar{\mu} = \frac{\int_{-\infty}^{\infty} n(x)x \, dx}{N} \qquad \text{or} \qquad \bar{\mu} = \frac{\sum_i n_i x_i}{N} \tag{5-6}$$

2. *Standard deviation or mean-square deviation.*

$$\sigma = \sqrt{\frac{\int_{-\infty}^{\infty} n(x)(\bar{\mu} - x)^2}{N}} \qquad \text{or} \qquad \sigma = \sqrt{\frac{\sum_i n_i(\bar{\mu} - x_i)^2}{N}} \tag{5-7}$$

Several other distribution parameters are occasionally used. Notice that the arithmetical average, or the *mean*, is determined in the most common way for an average. All the values are simply added and the sum is divided by the number of values. The *median* is the value such that half the readings are greater and half are lower; on a distribution graph, half the area is to the right of the median and half to the left. For instance, the mean of (1, 2, 3, 6, 8) is 4, but the median is 3. The *mode* is the most common value, the value at the peak of the distribution curve. With a Gaussian distribution, or any symmetrical distribution with a central peak, all three averages have the same value.

There are also several other measures of the spread of the distribution. The *variance* is the square of the standard deviation—just σ^2, or $\sum n_i(\bar{\mu} - x_i)^2/N$. The *average deviation* is defined as

$$\overline{\Delta x} = \frac{\int_{-\infty}^{\infty} f(x)|x| \, dx}{N} \qquad \text{or} \qquad \overline{\Delta x} = \frac{\sum_i n_i |\bar{\mu} - x_i|}{N} \tag{5-8}$$

For a Gaussian distribution, $\overline{\Delta x} = \sigma/1.25$. The probable error, p, is the width on either side of the mean such that half the distribution lies within $\bar{\mu} - p$ and $\bar{\mu} + p$. For a Gaussian distribution, $p \approx \sigma/1.5$.

$$\int_{\mu - \overline{\Delta x}}^{\mu + \overline{\Delta x}} \frac{N}{\sigma\sqrt{2\pi}} e^{-(\mu - x)^2/2\sigma^2} dx = 0.57N \qquad \begin{array}{l} \text{(area within} \\ \text{average deviations)} \end{array} \tag{5-9}$$

$$\int_{\bar{\mu}-p}^{\bar{\mu}+p} \frac{N}{\sigma\sqrt{2\pi}} e^{-(\bar{\mu}-x)^2/2\sigma^2} dx \quad = 0.5N \qquad \text{(area within probable errors)} \qquad (5\text{–}10)$$

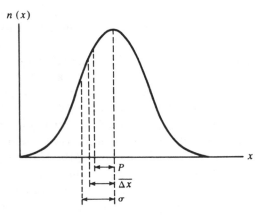

In our definitions of the mean and the standard deviation, we have been assuming that we are dealing with the entire population of data. Usually, we must be content with a sample. In the literature of statistics the sample parameters are distinguished from the population parameters by using English letters instead of Greek letters. The population mean is $\bar{\mu}$ and the corresponding standard deviation is σ. For the sample, the mean is called $\bar{x}$, and the standard deviation is s. There is one other difference that is important only for small samples. In determining the standard deviation of a sample, the mean has to be determined by adding all the values obtained and then dividing by the total number of values. In the process the number of independent variables has been reduced by 1, since given the mean, $\bar{x}$, and $n - 1$ of the values, the nth value could be calculated. The technical explanation is that we have lost a *degree of freedom*. Therefore, the squared deviations are averaged by dividing by $n - 1$ instead of by n. The formula for the standard deviation of the sample is

$$s = \sqrt{\frac{\sum_i (\bar{x} - x_i)^2}{n - 1}} \qquad (5\text{–}11)$$

Many math tables give values for the probability integral defined in this way: $2/\sqrt{\pi} \int_0^y e^{-y^2} dy$. In this form the quantity inside the integral is symmetric about $y = 0$. The integral could be taken from $-y$ to $+y$, which would eliminate the factor of 2. The relationship between this integral and the Gaussian form that we have given is that $y = (\bar{\mu} - x)/\sqrt{2}\sigma$. As an example of how to make this conversion from the form convenient for the data to the form necessary for the tables, let us find the value of the Gaussian integral from $-\sigma$ to $+\sigma$. This means taking the probability integral from 0 to $1/\sqrt{2}$, because if $\bar{\mu} - x = \sigma$, then $y = 1/\sqrt{2}$. The integral converts as follows:

5.1 DEFINITIONS AND PROPERTIES OF THE GAUSSIAN

$$\frac{1}{\sqrt{2\pi}\sigma} \int_{\mu-\sigma}^{\mu+\sigma} e^{-(\mu-x)^2/2\sigma^2} dx \longrightarrow \frac{2}{\sqrt{\pi}} \int_0^{1/\sqrt{2}} e^{-y^2} dy \qquad (5\text{--}12)$$

For $y = 1/\sqrt{2} = 0.71$, the table value for the probability integral is 0.68. Thus 68 per cent (or about two thirds) of the area of the integral lies between $-\sigma$ and $+\sigma$.

The derivation of the Gaussian function is given in the references cited at the end of this chapter. The assumptions that lead to such an exponential form are surprisingly few. It is assumed that the total probability of finding an event in some region of the random variable lying between $-\infty$ and $+\infty$ must be 1. Second, it is assumed that the maximum of the distribution is at the mean. Finally, it is assumed that the standard probability law holds: the probability of finding an event first in one interval *and* then in a second is equal to the *product* of the probabilities of finding the event in each interval separately. This assumption also implies that the individual readings are independent of each other—that the finding of one event does not prohibit or enhance the finding of any other event. An event, for instance, might be the measurement of an IQ score of one particular student.

It is astonishing that such general assumptions must lead to such a unique mathematical equation. It is even more curious that so many distributions are either strictly Gaussian or close approximations to it. In Section 4.6 we show that the binomial distribution approaches the Gaussian function for large sample values. The fact that the distribution of numbers in dice throws is approximately Gaussian can be derived from obvious considerations of the probability conditions on each die. It is not so obvious why biological characteristics should frequently be Gaussian. One might think, for example, that IQ scores should be skewed toward the high end because humans with low intelligence would have a low survival rate. To explain the fact that the IQ distribution is Gaussian, we might challenge the validity of the test, or question its significance during past generations. A more likely explanation is that the necessary probability requirements are apt to be met whenever a property is determined by many different factors working independently. Human intelligence, or whatever is tested by the IQ marks, is not the result of the action of a single gene, but depends on a great many factors.

Precise measured values for any quantity determined by experiment usually cluster in a Gaussian distribution around some central value. The uncertainty, or the error, is usually caused by a number of factors acting independently and subject to the probability conditions that yield the Gaussian distribution.

C. As an example of the problem of using statistical theory to fit experimental data, consider the following data:

55 readings taken of
period of vibrating reed

t (sec)	Number of readings in bin
0.015	1
0.016	3
0.017	3
0.018	6
0.019	9
0.020	11
0.021	8
0.022	7
0.023	4
0.024	2
0.025	0
0.026	1

$n = 55$

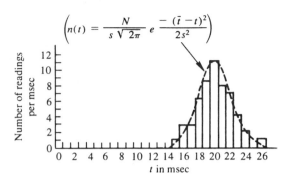

$$\left(n(t) = \frac{N}{s\sqrt{2\pi}}\, e^{\dfrac{-(\bar{t}-t)^2}{2s^2}} \right)$$

Some timing device apparently yielded information in millisecond intervals or else an observer judged its readings within those intervals. The properties of either the vibrating reed or the measuring system, including the observer, gave rise to the spread of data shown. Since the distribution is fitted fairly well by a Gaussian curve, $n(t) \propto e^{-(\bar{t}-t)^2/2s^2}$, it is reasonable to apply the statistical parameters. [This fit was actually assumed in using the root-mean-square summation to find the standard deviation. The standard deviation was then used to calculate the points for the curve given by $n(t)$. Since the curve fits, the original assumption was justified.]

There may be reasons not to use the theory: (1) Suppose that the timing device could not read lower than 18 ms. The original distribution would be obviously skewed. The formulas for $\bar{t}$ and s would yield values, but it would be meaningless to use them. (2) Suppose that only 10 readings had been taken instead of 55. The fluctuations that are apparent with the 55 readings would be even greater, so that the distribution could hardly be Gaussian and the use of the formal apparatus would allow only specially qualified predictions. (3) Most of the data may fit the Gaussian distribution, but random mistakes might produce an occasional very high or very low value. The experimenter then must decide whether these strange values are to be thrown out or are to be studied further as possible clues to new phenomena. (Several Nobel prizes have been lost by making the wrong decision.) If a measured value is much greater than $\bar{x} + s$ or much smaller than $\bar{x} - s$ (and if the other data is sufficient to justify the use of the Gaussian function), some suspicion is justified that the reading is a mistake. On the other hand, abnormally high or low values are to be expected occasionally; indeed, if about one third of the events are not outside $\pm s$, there must be something wrong with the data or the method. (4) The vibrating reed might be driven by some voltage-sensitive device that would respond to periodic swings of line voltage by encouraging vibration periods close to the upper and lower ends of the reed's range, suppressing periods near the natural period of the reed. The data distribution would then not look at all Gaussian and the standard error analysis should not be used. (5) Finally, the data might match a Gaussian curve exactly, and the scatter of points, measured by s, might be extremely small. But the timer might throw in three extra milliseconds to every reading, and this would make the accuracy poor and the elaborate analysis futile.

5.2 SAMPLING PROBLEMS. Under most circumstances it is hard or impossible to deal with a whole population of data. Instead we analyze a sample. There are some obvious and nonmathematical restrictions on the choice of a sample. In many social science applications these restrictions are formidable. A sample for a voters' poll, for instance, should be proportionately representative of various geographical regions, various religions, various age groups, various ethnic groups, etc. The proportionality must be weighted by realities of electoral effect. A famous poll comprised of all the middle-class citizens who subscribed to a particular magazine and who answered a questionnaire before the 1936 election predicted that Landon would win the presidency in a landslide. The inaccuracy resulting from this means of sampling became apparent when Landon carried only 2 of 48 states. Our present poll takers are more sophisticated and more successful.

Mathematical theory can deal with certain sampling problems providing that the samples are chosen at random or without prejudice from a large population, preferably one characterized by a Gaussian distribution. If you choose a large enough sample, you might expect that its mean, $\bar{x}$, would be close to the population mean, $\bar{\mu}$. The problem is, how large a sample do you need to feel confident that $\bar{x}$ is within a certain range of $\bar{\mu}$? Each new sample that you draw will probably have a different mean. If you plot the values for these means of the samples, you will get a new distribution. This distribution is sharper than the population distribution. As the number of samples increases, the mean of this distribution of the sample means approaches the population mean. Furthermore, according to the Central Limit Theorem, as the number, n, of items in each sample increases, the distribution of the means approaches a Gaussian function with a standard deviation of $\sigma/\sqrt{n}$, where σ is the standard deviation of the population. This powerful Central Limit Theorem holds even under many conditions when the population distribution itself is not Gaussian. The content of the theorem is summarized in the diagrams on page 93. The remarkable feature is that the distribution of the sample means is a very specific distribution—a Gaussian function for large enough sample size. If the population distribution is Gaussian, then the distribution of the means is Gaussian for *any* size sample.

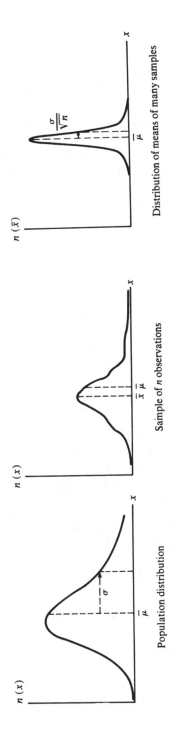

Population distribution

Sample of n observations

Distribution of means of many samples

93

5.3 IDENTIFICATION TEST FOR LARGE SAMPLE—u TEST. Two problems might arise concerning a sample of data. First, if you know that your sample is taken from a particular parent population, you should know how close the sample mean, $\bar{x}$, may be to the unknown population mean, $\bar{\mu}$. Second, it may be that the problem is whether or not the sample is indeed part of a particular population. For instance, the mean IQ for a "standard" population in the United States is set at 100 with a standard deviation of 16. If a group of 100 students has a mean IQ of 105, are they a specially chosen group or could they be a random sample from the population? Note that there is no question but that according to the test the group has above average intelligence. The question is whether or not this might be attributed to a sampling fluctuation.

The Central Limit Theorem says that the distribution of the sample means is Gaussian with a standard deviation of $\sigma/\sqrt{n}$, where σ is the standard deviation of the population and n is the number of items in each sample. For large samples, the sample standard deviation, s, is approximately the same as the population standard deviation, σ. Suppose that we want to guarantee that the population mean is close to the sample mean within certain limits 95 per cent of the time. The assumption is that we might take endless numbers of samples of n items each, in which case 95 per cent of their means would fall in the region between

$$\bar{\mu} - \frac{1.96\sigma}{\sqrt{n}} \quad \text{and} \quad \bar{\mu} + \frac{1.96\sigma}{\sqrt{n}}.$$

The shaded region in the diagram shows the 95 per cent of the area of the Gaussian function that is enclosed between these two limits. Other probability limits are listed in the accompanying table. The horizontal bars on the graph represent various samples that might be taken, each having a slightly different mean, $\bar{x}$, but with approximately the same standard deviation of s. If only statistical fluctuations are responsible for the variation in sample means, we should expect that 95 times out of 100 the population mean, $\bar{\mu}$, will be included within a sample mean $\pm 1.96\, s/\sqrt{n}$.

$$\text{probability}\left[\left(\bar{x} - \frac{1.96\,s}{\sqrt{n}}\right) < \bar{\mu} < \left(\bar{x} + \frac{1.96\,s}{\sqrt{n}}\right)\right] = 0.95$$

The "confidence interval" is the region between

$$\bar{x} - \frac{1.96\,s}{\sqrt{n}} \quad \text{and} \quad \bar{x} + \frac{1.96\,s}{\sqrt{n}}$$

and the probability is called the *confidence level*. Sometimes a confidence level of 95 per cent is described as an α level of 0.05. Note that there is a 2.5 per cent chance that $\bar{\mu}$ will be greater than $\bar{x} + 1.96\, s/\sqrt{n}$ and a 2.5 per cent chance that $\bar{\mu}$ will be less than $\bar{x} - 1.96\, s/\sqrt{n}$.

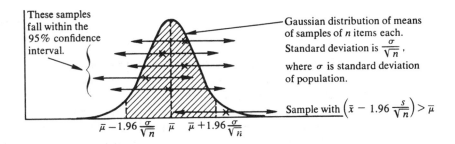

These samples fall within the 95% confidence interval.

Gaussian distribution of means of samples of n items each. Standard deviation is $\frac{\sigma}{\sqrt{n}}$, where σ is standard deviation of population.

Sample with $\left(\bar{x} - 1.96\frac{s}{\sqrt{n}}\right) > \bar{\mu}$

$\bar{\mu} - 1.96\frac{\sigma}{\sqrt{n}}$ $\bar{\mu}$ $\bar{\mu} + 1.96\frac{\sigma}{\sqrt{n}}$

Fraction of area	Confidence interval
0.50	$\bar{\mu} \pm 0.675\frac{\sigma}{\sqrt{n}}$
0.80	$\bar{\mu} \pm 1.282\frac{\sigma}{\sqrt{n}}$
0.90	$\bar{\mu} \pm 1.645\frac{\sigma}{\sqrt{n}}$
0.95	$\bar{\mu} \pm 1.960\frac{\sigma}{\sqrt{n}}$
0.98	$\bar{\mu} \pm 2.326\frac{\sigma}{\sqrt{n}}$
0.99	$\bar{\mu} \pm 2.576\frac{\sigma}{\sqrt{n}}$
0.999	$\bar{\mu} \pm 3.291\frac{\sigma}{\sqrt{n}}$

Areas enclosed within confidence interval of Gaussian with standard deviation of $\frac{\sigma}{\sqrt{n}}$

As for whether or not there is something special about the group of 100 students with a mean IQ of 105, note that there would be nothing at all unusual about picking one student at random and discovering that his IQ is 105. That score is well within one standard deviation (16) of the population mean. If many samples of 100 each are chosen, however, their means will form a Gaussian distribution with a mean of 100 and a standard deviation of

$$\frac{\sigma}{\sqrt{n}} = \frac{16}{\sqrt{100}} = 1.6.$$

The score of 105 is a little over three of these standard deviations from the mean. There is only one chance in a thousand that a group of 100 chosen at random from the general population would have an IQ score less than 95 or greater than 105. Such a possibility could occur, of course (and ought to on the average of once every thousand times), but the chances are that there was some selection process at work in assembling that group of students.

This analysis of the location of the sample mean is sometimes called the *u test* for the following reason. A parameter, *u*, is defined

$$u = \frac{\bar{x} - \bar{\mu}}{\sigma/\sqrt{n}} \qquad\qquad (5\text{--}13)$$

This parameter is the ratio of the spacing (between the sample mean and the population mean) to the standard deviation (of the distribution of means of true samples). About two thirds of the means of true samples will satisfy $|u| < 1$. About 95 per cent will satisfy $|u| = < 1.96$. ($|u|$ means the absolute value of u without regard to sign.) Consequently, if the $|u|$ for any test sample is greater than 1.96, it represents an unusual occurrence with a probability of only 5 per cent. There is therefore high probability that the sample was not a member of the assumed population. Note that there is no new content in this u test; it is just a rephrasing of the previous analysis.

5.4 IDENTIFICATION TEST FOR SMALL SAMPLES—STUDENT'S t TEST.
If the standard deviation of the population is not known, it can be approximated by finding the standard deviation of a *large* sample. For small samples, however, s is not a good measure of σ. Since σ might be larger or smaller than s, the distribution of the sample means may spread out with a larger standard deviation than $s/\sqrt{n}$. The smaller the sample size, n, the greater the probability that σ might be larger than s, and therefore the greater the probability that the distribution of $\bar{x}$ is wider than expected. An analytical solution to this problem was worked out at the turn of this century by an Irish chemist named William Gosset. His employers did not want him to publish under his own name, and so he used the pseudonym "Student." The criterion that he devised for this small-sample problem is known as the t test, or Student's t test.

Note that the actual distribution of means, even for small samples, is Gaussian (if the population is Gaussian), and the actual standard deviation is $\sigma/\sqrt{n}$. The trouble is that $s/\sqrt{n}$ derived from a particular small sample gives an uncertain approximation to $\sigma/\sqrt{n}$. Student's (Gosset's) solution was to define a parameter, t, similar to u of the u test, but which contains a correction factor dependent on the size of the sample.

$$t = \frac{\bar{x} - \bar{\mu}}{s/\sqrt{n}} \qquad\qquad (5\text{--}14)$$

Usually, t is given in the form of tables containing columns for several confidence levels and a separate row for each sample size. Instead of listing sample size, n, the "degrees of freedom" are given. In most cases, the number of degrees of freedom is 1 less than the number n in the sample. One degree of freedom is lost in taking the mean in order to determine s for the sample. Here is a short table of t values.

Degrees of Freedom	t for Various Confidence Intervals			
	0.50	0.90	0.95	0.99
1	1.000	6.314	12.706	63.657
2	0.816	2.920	4.303	9.925
3	0.765	2.353	3.182	5.841
4	0.741	2.132	2.776	4.604
5	0.727	2.015	2.571	4.032
6	0.718	1.943	2.447	3.707
7	0.711	1.895	2.365	3.499
15	0.691	1.753	2.131	2.947
30	0.683	1.697	2.042	2.750
99	0.676	1.660	1.984	2.626
∞	0.674	1.645	1.960	2.576

For infinite degrees of freedom, $t = u$. To determine whether or not a sample is probably a member of a population, calculate its t value. For instance, suppose that a sample of 6 items is drawn with the values 8, 4, 7, 9, 6, and 8.

$$
\begin{array}{cc}
 & (x - \bar{x}) & (x - \bar{x})^2 \\
8 & 1 & 1 \\
4 & 3 & 9 \\
7 & 0 & 0 \\
9 & 2 & 4 \\
6 & 1 & 1 \\
8 & 1 & 1 \\
\hline
6\,\lfloor 42 & & 5\,\lfloor\overline{16} \\
\overline{7 = \bar{x}} & & \overline{3.2}
\end{array}
$$

$$ s = \sqrt{\frac{(x - \bar{x})^2}{n - 1}} = \sqrt{3.2} = 1.79 $$

$$ t = \frac{\bar{x} - \bar{\mu}}{s/\sqrt{n}} = \frac{7 - \bar{\mu}}{1.79/\sqrt{6}} = \frac{7 - \bar{\mu}}{0.73} $$

If it is known that the population mean is $\bar{\mu} = 9$, then the sample $t = 2.74$. For 5 degrees of freedom $(n - 1)$, that large a t value would occur for only about 3 per cent of true samples of the population. Ninety five per cent of such true samples will have $t < 2.57$, although 99 per cent will have $t < 4.03$.

The distribution of u values is Gaussian centered on $u = 0$ with a standard deviation of 1.0. The distribution of t values obtained from repeated sampling forms a bell-shaped curve with a larger spread than the u distribution. The smaller the sample size, the larger the spread of the t distribution. Several of these distributions are shown in the diagram on p. 262.

Sometimes statisticians refer to "one-tail" tests or "two-tail" tests. This is a reference to the regions outside the confidence intervals of the distribution curves. There is a 5 per cent probability that a large true sample of a population will have a u value greater than $+1.96$ *or* less than -1.96. In other words, there is a 5 per cent chance that the u value of the sample will fall in one of the two tails of the distribution. Each tail will contain $2\frac{1}{2}$ per cent of all possible samples. If you are concerned that your sample be within the 95 per cent confidence level *above* the mean, then its u value should be less than 1.645. Then 5 per cent of the true samples might have u larger than this and be represented in the upper tail of the distribution. Note that the value 1.645 is the u value for the 90 per cent confidence level—5 per cent in the upper tail and 5 per cent in the lower tail.

5.5 COMPARISON OF SAMPLE VARIANCE—χ^2 TEST.

A sample drawn from a Gaussian (sometimes called "normal") population will not generally have the same standard deviation as that of the population. The standard deviation of a very large sample (100 or more) will probably be about the same as that of the population, but fluctuations should be expected. The actual values of standard deviations of many samples could be plotted and would form a distribution about the value for the population. A particular form of this distribution is usually used making use of the variances σ^2 and s^2. The variance, σ^2, is a constant of the population, but s^2 changes from sample to sample. We "normalize" the sample variance, s^2, by dividing it by the population variance, σ^2. The criterion formed is given a special name, *chi square:*

$$\chi^2 = (n - 1)\frac{s^2}{\sigma^2} \tag{5-15}$$

The distribution graphs for χ^2 for various values of sample size, n, are shown in the diagram on p. 260. Note that for very large values of n, the χ^2 distribution is centered approximately around n, since for large samples $s^2 \approx \sigma^2$.

The χ^2 test could be used to determine whether or not a sample is probably a true member of some known population. Usually, however, the u or t test is sufficient for such a purpose. The χ^2 test is more generally used to compare an observed frequency distribution with an expected one. The distributions to be compared can have any form and, in particular, need not be Gaussian. It is the *difference* between each experimental value and the expected value that becomes a point on a distribution curve, and it is the variance of *this* curve that is then analyzed in terms of a χ^2 distribution.

Here is an illustration of a typical problem of this type. Suppose that data can be fitted into k different ranges or "bins." The bins might, for instance, represent the scattering angles for a particular type of particle collision. Allot 10 bins to cover the range 0 to 180°, with each bin containing the data falling into that particular 18° interval. If the probability of a scattering falling into the ith bin is p_i, then the expected number in that bin is np_i, where n is the total number of scatterings. The bin index, i, goes from 1 to k (in this case, 1 to 10) and $\sum_{i=1}^{k} p_i = 1$. (Each scattering must land in *one* of the bins.) For any particular experimental run there will be fluctuations around the expected numbers in each bin. The standard deviation of the distribution in the ith bin, which would be found from compiling many experimental runs, is $\sigma_i = \sqrt{np_i}$. (In the section on the binomial distribution we showed that $\sigma = \sqrt{npq}$ when the binomial can be approximated by a Gaussian. But $q = (1-p) \approx 1$, since p_i is small for these error distributions. This situation also applies for a multinomial distribution, such as the one in our example.) The diagram illustrates a hypothetical example of scattering data.

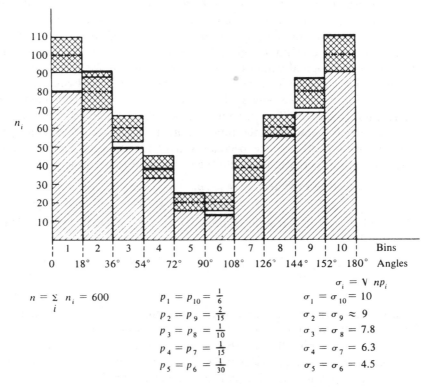

$$n = \sum_i n_i = 600$$

$p_1 = p_{10} = \frac{1}{6}$

$p_2 = p_9 = \frac{2}{15}$

$p_3 = p_8 = \frac{1}{10}$

$p_4 = p_7 = \frac{1}{15}$

$p_5 = p_6 = \frac{1}{30}$

$\sigma_i = \sqrt{np_i}$

$\sigma_1 = \sigma_{10} = 10$

$\sigma_2 = \sigma_9 \approx 9$

$\sigma_3 = \sigma_8 = 7.8$

$\sigma_4 = \sigma_7 = 6.3$

$\sigma_5 = \sigma_6 = 4.5$

The expected values and standard deviations are shown by dotted lines and crosshatched areas. The actual values of one experimental run are shown by solid lines and shaded regions.

i	Observed	Expected	$|O - E|$	$(O - E)^2$	$(O - E)^2/E$
1	80	100	20	400	4.00
2	90	80	10	100	1.25
3	50	60	10	100	1.67
4	40	40	0	0	0
5	25	20	5	25	1.25

i	Observed	Expected	$\mid O - E \mid$	$(O - E)^2$	$(O - E)^2/E$
6	15	20	5	25	1.25
7	45	40	5	25	0.62
8	55	60	5	25	0.42
9	90	80	10	100	1.25
10	110	100	10	100	1.00
					12.71

The question is, how serious are the fluctuations of the observed distribution when compared with the fluctuations expected from chance? Is the data a true sample of the population? A measure of the expected fluctuations in s compared with σ is given by the χ^2 distribution. Our problem is to transform the original definition of χ^2 to one suitable for this type of case.

Here the *fluctuations* in the bins from one experimental run form a distribution for which we would like to know s^2. Each of these fluctuations represents a point of that distribution, but each should be normalized to compensate for the variable number, n_i, in that bin. The appropriate normalization factor is the expected standard deviation, σ_i, for that bin. Our distribution of fluctuations should then consist of these values:

$$\frac{\text{observed}_i - \text{expected}_i}{\sigma_i} = \frac{O_i - E_i}{\sqrt{np_i}} = \frac{O_i - E_i}{\sqrt{E_i}}$$

(Remember that in bin i the expected value, E_i, is equal to np_i.) The χ^2 function becomes

$$\chi^2 = (n - 1)\frac{s^2}{\sigma^2} \longrightarrow \sum_{i=1}^{k} (k - 1)\frac{(O_i - E_i)^2/(k - 1)}{\sigma_i^2} = \sum_{i=1}^{k} \frac{(O_i - E_i)^2}{E_i}$$

$$(5\text{--}16)$$

The n in the original definition becomes k, the number of bins, since each bin contributes one point to the s^2 value. For small number of bins and small number in each bin, this expression only gives an approximation to the theoretical χ^2. As a practical matter, the approximation is usually satisfactory if both k and E are equal to or greater than 5.

We have already performed the operations to find χ^2 for our example of a scattering distribution. The table shows that the χ^2 value is 12.71. This should be compared with the distribution of χ^2 expected from chance. Examine the χ^2 curve for 9 degrees of freedom ($k - 1$), and observe that the value 12.71 comes at a point where almost 18 per cent of the area is in the tail. In other words, you would expect that chance alone would produce χ^2 values greater than 12.7 about 18 per cent of the time. Therefore, at the 18 per cent significance level the variability in counts is not excessive.

Let us take another example. Suppose that you obtain radioactive emission counts on each of 10 small squares cut from a plastic that is supposed to have been uniformly irradiated. The counts from the 10 samples are 141, 117, 107, 134, 110, 149, 107, 152, 122, and 131. The mean is 127 and that can be used as the expected count (E). In doing this, however, we sacrifice another degree of freedom. [The first was lost in choosing a fixed number (10) of samples.] Is it possible that the variability in counts is greater than can be accounted for by chance?

The readings are plotted in the diagram, showing a histogram with interval ranges of 5 counts. The expected standard deviation is equal to $\sqrt{127}$, or 11 counts. A Gaussian curve centered at 127 with a standard deviation of 11 and appropriate area is sketched in the diagram. It is not improbable

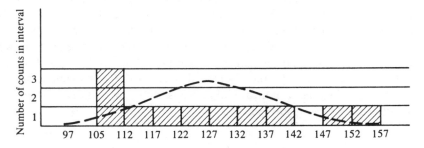

to find any particular reading, although the count of 152 is a little unlikely. To get a measure of the probability of obtaining a *group* of readings like this, calculate the experimental value of χ^2.

$$\chi^2 = \frac{(141 - 127)^2}{127} + \frac{(117 - 127)^2}{127} + \frac{(107 - 127)^2}{127} + \frac{(134 - 127)^2}{127}$$
$$+ \frac{(110 - 127)^2}{127} + \frac{(149 - 127)^2}{127} + \frac{(107 - 127)^2}{127} + \frac{(152 - 127)^2}{127}$$
$$+ \frac{(122 - 127)^2}{127} + \frac{(131 - 127)^2}{127} = 20.15$$

The table summarizes χ^2 values for a variety of different degrees of freedom and for various confidence levels. According to that table, for 8 degrees of freedom only 1 per cent of chance fluctuations will produce $\chi^2 \geqq 20.09$. We could accept the hypothesis that the samples were uniformly irradiated only at the 1 per cent significance level. Probably the samples were not uniformly radioactive.

5.6 CORRELATION OF TWO VARIABLES—THE r COEFFICIENT. How do you demonstrate that two variables are related to each other? Usually we look first for a theoretical model that calls for such a relationship. The period of a pendulum is proportional to the square root of the length of the string, and this can be derived from basic considerations of the dynamics. Alternatively, you could measure the periods produced by various lengths and plot $T(L)$. You would get a curved line and then might try graphing

T versus L^2. If the experimental points were carefully taken, they would fall smoothly on a straight line. Of course, there would be some scatter around the line because of experimental error.

If experimental points subject to large error are obtained, it may not be so obvious that there is any relationship between two sets of variables. In Section 3.J formulas were given for obtaining a best-fit straight line through scattered data. This line is called the *line of regression of y on x* and is designed to minimize the sum of the squared deviations of y from the line.

In the diagram we sketch two sets of data that are satisfied by the same line of regression. Clearly, however, there is much greater correlation between the two variables in the first graph than in the second. The dispersion shown in the second case could be caused by large experimental errors, or it might be that the actual causal relationship between the two variables is weak. It might be necessary, for instance, to have a lot of rain to get a good rice crop, but there may be many other factors also at work. Consequently, high rain levels might in general indicate large crops, and low rain levels would yield small crops, but not always.

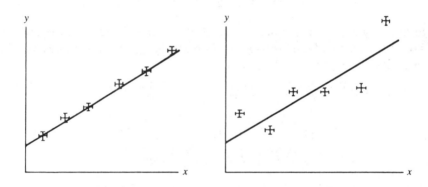

A standard coefficient of correlation, r^2, has been defined to characterize the dispersion of data around the *best-fit* or *least-squares* line. It is very important to note that the use of this coefficient depends on the assumption that x and y are linearly related. If $x = \sin y$ or if $x^2 + y^2 = k$ (a circle), the variables would be highly correlated, but a value of r^2 might indicate no correlation at all for the (x, y) relationship (There is, however, a linear relationship between x and $\sin y$ in the first case and x and $\sqrt{k - y^2}$ in the second.)

Consider the squared deviations from the two different lines of the points shown in the next diagram. The horizontal line goes through the mean of the y values. So does the line of regression. If there is linear correlation so that the points cluster along the line of regression, the squared deviations from that line should be less than those from the horizontal line. The coefficient of correlation is defined as the difference between the sums of those

102

sets of deviation, normalized by the deviations from the horizontal. In that way, the coefficient is a pure number.

$$r^2 = \frac{\sum (y_i - \bar{y})^2 - \sum [y_i - (b + mx_i)]^2}{\sum (y_i - \bar{y})^2} \qquad (5\text{--}17)$$

The first term in the numerator is the sum of the squared deviations of the points from a horizontal line passing through the mean, $\bar{y}$. The second term is the sum of the squared deviations of the points from the line of regression, $y = b + mx$. If the data points fall exactly on the line of regression, there is perfect correlation, the second term is zero, and $r^2 = 1$. The maximum value that the second term can have is equal to $\sum (y_i - \bar{y})^2$. In that case the squared deviations from the line of regression are just as great as they are from the horizontal line. There is no linear correlation, the numerator is zero, and therefore $r^2 = 0$. The dashed vertical lines indicate $y_i - \bar{y}$. The solid vertical lines represent $y_i - (b + mx_i)$.

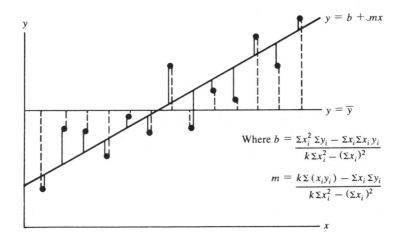

$$\text{Where } b = \frac{\sum x_i^2 \, \sum y_i - \sum x_i \sum x_i y_i}{k \sum x_i^2 - (\sum x_i)^2}$$

$$m = \frac{k \sum (x_i y_i) - \sum x_i \sum y_i}{k \sum x_i^2 - (\sum x_i)^2}$$

To find r^2 from data, the equations for m and b must be substituted into the original equation. Algebraic manipulation yields a simple formula for r, instead of r^2.

$$r = \frac{n \sum x_i \, y_i - (\sum x_i)(\sum y_i)}{\sqrt{[n \sum x_i^2 - (\sum x_i)^2][n \sum y_i^2 - (\sum y_i)^2]}} \qquad (5\text{--}18)$$

or

$$r = \frac{\sum x_i y_i - n\bar{x}\bar{y}}{n\sigma_x \sigma_y} = \frac{\sum (x_i - \bar{x})(y_i - \bar{y})}{n\sigma_x \sigma_y}$$

This last expression yields another geometric interpretation of r. Notice that the deviations to be multiplied in the numerator are those from the horizontal and vertical lines through the means of the scattered data. As shown in the next diagram, this corresponds to a shift in axes. Furthermore, the deviations have polarity, $+$ and $-$; they are not squared. Consequently, their products can be $+$ or $-$. If a data point is in the first or third quadrant of the transformed plot with the shifted axes, the product of its deviations from the axes is positive. If the data point is in the second or fourth quadrant, its product of deviations is negative.

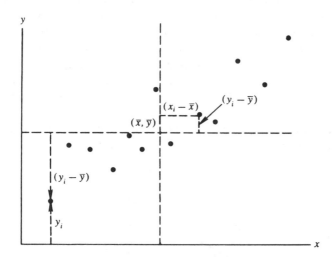

In the final formula for r, note that the standard deviations in the denominator serve to normalize the deviations in the numerator, thus making them independent of scale or units. The n in the denominator serves to give an average of the products of the deviations, and so makes r independent of the number of data points. If there is a high positive correlation, the data points will fall mostly in the third and first quadrants, producing positive r with a maximum value of $+1$. High negative correlation would be represented by data points mostly in the second and fourth quadrants, producing negative r with a maximum negative value of -1. Both cases would yield r^2 close to 1, and so signify linear correlation. In the negative case, the slope of the line of regression is negative. If the points are dispersed throughout all the quadrants, the positive products will cancel the negative products, producing a low value of r, implying low correlation.

There are two common fallacies in correlation analysis. One is forgetting that the simple formulas are concerned with a linear relationship between the two variables. The second is assuming causal relationships between variables just because r^2 is closer to 1 than to 0. As an example of this latter problem, consider these observations made during the course of a physics lecture. Every 5 minutes an observer made a subjective judgment of the number of students who were obviously awake and following the lecture. The lecture was taped and afterwards a count was made at the end of each 5 minutes of the number of words per minute spoken by the lecturer. The data is presented in the table and graphed in a "scatter diagram."

Time	No. Alert Students (y)	No. Words/min (x)	xy	x^2	y^2
0	100	125	12,500	15,600	10,000
5	95	125	11,900	15,600	9,000
10	90	130	11,700	16,900	8,100
15	80	140	11,200	19,600	6,400
20	60	150	9,000	22,500	3,600
25	55	155	8,500	24,000	3,000
30	55	165	9,100	27,200	3,000
35	50	180	9,000	32,400	2,500
40	50	190	9,500	36,100	2,500
45	40	200	8,000	40,000	1,600
50	30	200	6,000	40,000	900
Sums	705	1,760	106,400	289,900	50,600

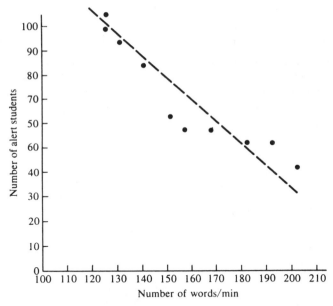

Number of words/min

Formula 5-18 is the easiest to use with most primitive data. Substituting the results from the table into the formula for r, we get

$$r = \frac{n \sum x_i y_i - (\sum x_i)(\sum y_i)}{\sqrt{[n \sum x_i^2 - (\sum x_i)^2][n \sum y_i^2 - (\sum y_i)^2]}}$$

$$= \frac{11(106,400) - (1760)(705)}{\sqrt{[11(289,900) - (1760)^2][11(50,600) - (705)^2]}}$$

$$= -0.95$$

$$r^2 = 0.9$$

5.6 CORRELATION OF TWO VARIABLES—THE r COEFFICIENT

We have demonstrated that there is very high correlation between the rate of lecturing and the attention of the audience—the faster the speaker, the less attentive the listeners. But is there a causal relationship? It is not unreasonable that the speaker lost part of his audience for this reason, but other factors are of obvious importance. Part of the audience probably tuned out simply as a function of their attention span. The hour wore on! As the lecturer sped up he probably bypassed important side explanations. Perhaps he was getting into more difficult material. The only sure conclusion is that the data collection and analysis—by itself—proved nothing.

5.7 CONCLUDING COMMENTS ABOUT STATISTICAL ANALYSIS. In this resume we have touched on only a few of the techniques of statistics, although these are the ones most frequently used. The proper use of statistical methods often requires great mathematical sophistication. For some insight into the problems and possibilities we refer you to the following introductory texts:

Introduction to Statistical Reasoning, P. J. McCarthy (McGraw-Hill, New York, 1957).
Introduction to Mathematical Statistics, P. G. Hoel (Wiley, New York, 1962).

The most useful technique for dealing with statistics is to graph the data. Tables and numerical coefficients can often be opaque or deceptive. It is a rare problem in correlation, for instance, in which better judgment can be drawn from a knowledge of r^2 than can be obtained by examining a scatter diagram of the data.

SAMPLES

1. Check the values for $\bar{t}$ and s given in the data about the vibrating reed. Using that value of s, check the points on the Gaussian distribution curve by evaluating $n(t)$. For instance,

$$n(t) = \frac{N}{s\sqrt{2\pi}} e^{-\frac{(\bar{t}-t)^2}{2s^2}}$$

For $(\bar{t} - t) = 2$ ms,

$$n(22) = \frac{55}{2.2\sqrt{2\pi}} e^{-(2)^2/2(2.2)^2} = 9.9e^{-0.41} = 6.6$$

2. Evaluate the error integral between $-p$ and $+p$.
3. Cast 10 coins 10 times, keeping track of the number of heads each cast. Plot a frequency curve against the number of heads possible (0–10). Employ the standard procedures to compute the average and the standard deviation. Compare the distribution graph with a standard Gaussian curve and observe to what extent the computed average and standard deviation fulfill their expected roles.

Build up the data with 100 throws, go through the same procedures, and answer the same questions.

4.   Most of the standard IQ tests are designed to have a mean of 100 for the general population and a standard deviation of 16. What percentage of the population would have an IQ between 116 and 132?   ($\sim 13\%$.) How many of the 200 million people in the United States have an IQ over 150? ($\sim 0.12\%$ or about 250,000)

5.   The number of counts/10 s recorded by a radiation detector is shown in the table. Plot a histogram (bar graph) of the data, compute the mean and the standard deviation, and then plot a Gaussian curve on the same graph for comparison.

							900–							
830	840	850	860	870	880	890	909	910	920	930	940	950	960	970	980
1	1	3	5	9	11	12	14	13	10	8	6	3	2	1	1

6.   Transform the defining formula for r^2 (5–17) to the forms for r given in (5–18).

7.   Two dice are thrown 720 times, yielding the following distribution:

2	3	4	5	6	7	8	9	10	11	12
34	35	47	63	87	173	93	70	44	38	36

Compute χ^2 for this distribution and decide whether or not the results were due to chance.

CHAPTER6 QUADRATIC AND HIGHER
POWER EQUATIONS

It often becomes necessary in solving problems to find the roots of simple algebraic equations. Although general solutions exist for equations up to the fourth degree, the solutions are so complicated that for equations higher than degree two, people generally turn to iteration methods applicable to high-speed computers (e.g., Newton's or Horner's method). The second-degree equation or *quadratic* has a simple solution with widespread applications and we shall go into it here.

6.1 QUADRATIC. A general quadratic equation has the form

$$y = ax^2 + bx + c \tag{6-1}$$

where a, b, and c are constants. The two roots of this equation are the values of x that make $y = ax^2 + bx + c = 0$, or where the curve of $y(x)$ crosses the x axis.

To derive a general solution, one which will work for any a, b, or c, we use a method called "completing the square" of the equation.

$$ax^2 + bx + c = 0 \qquad ax^2 + bx = -c \qquad x^2 + \frac{b}{a}x = -\frac{c}{a}$$

$$x^2 + \frac{b}{a}x + \frac{b^2}{4a^2} = -\frac{c}{a} + \frac{b^2}{4a^2}$$

$$\left(x + \frac{b}{2a}\right)^2 = \frac{b^2}{4a^2} - \frac{c}{a}$$

Taking the square root of both sides gives

$$x + \frac{b}{2a} = \pm\sqrt{\frac{b^2}{4a^2} - \frac{c}{a}} \tag{6-2}$$

Notice that both the $+$ and $-$ signs are needed—either one squared will produce the right-hand side above. Rearranging terms under the radical sign and subtracting $b/2a$ from both sides gives the standard form of the quadratic formula:

$$x = \frac{-b \pm \sqrt{b^2 - 4ac}}{2a} \tag{6-3}$$

We can try it out for simple cases just to make sure that we haven't made an arithmetic mistake. For instance, $y = (x + 2)(x - 1) = x^2 + x - 2$. $y = 0$ at $x = -2$ and $x = 1$. Putting the values $a = 1$, $b = 1$, and $c = -2$ into our formula gives

$$x = \frac{-1 \pm \sqrt{1 + 8}}{2} = \frac{-1 \pm \sqrt{9}}{2} = \frac{-1 \pm 3}{2} = -2, +1$$

The quadratic equation, as we show on p. 134, is just the general equation of a parabola, and the quadratic formula gives the x values where the parabola crosses the axis. If you look at the formula, you can see that there are three possibilities depending on the sign of $b^2 - 4ac$, the quantity under the radical, which is called the *discriminant*. If $b^2 - 4ac$ is positive, then the formula gives two distinct real roots. If the discriminant is 0, we have $x = -(b/2a) \pm 0 = -(b/2a)$ and the two roots are identical. This will occur if a maximum or a minimum point lies on the x axis, as is the case when $y = x^2$.

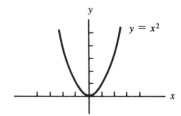

The third possibility occurs when $b^2 - 4ac < 0$ so that both roots are complex. This corresponds to a parabola that doesn't cross the x axis at all. For instance, consider the equation, $y = -(1 + x^2)$. For $y = 0$, $x = \pm i$.

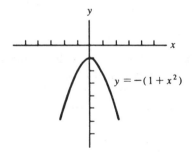

1. Show that the sum of the two roots of any quadratic equation satisfies the equation: $x_1 + x_2 = -b/a$. Similarly, show that the product $x_1x_2 = c/a$. Why can't we treat these as simultaneous equations and force out a solution by solving for x_1 or x_2?

Find the roots of

2. $y = x^2 + 3x + 2$, $x = -1, -2$

3. $y = 3x^2 + x - 5$, $x = \dfrac{-1 \pm \sqrt{61}}{6}$

4. $y = 2x^2 - x + 2$, $x = \dfrac{1 \pm \sqrt{-15}}{4} = \dfrac{1 \pm i\sqrt{15}}{4}$

5. $y = x^2 + 4x + 4$, $x = -2$

6.2 FACTORING EQUATIONS. There are general algebraic solutions to the cubic and quadric equations, although they are so complicated that it is usually easier to solve such equations by graphing them, or by guessing at approximations. There are no general algebraic solutions to equations of fifth or higher order. You can sometimes manage to find the roots of higher powered equations by factoring out terms. For example, if you were given the equation $y = x^3 - 2x^2 - 2x - 3$ and told to find the roots, you might after some grumbling be able to rewrite it as $y = (x - 3)(x^2 + x + 1)$. This is equivalent to the first equation as you can see by multiplying it out. The first factor is zero at $x = 3$, and therefore 3 must be a root of the original equation. The second factor is just a quadratic, which you already know how to solve.

The way to factor an equation is to try various values for x until you find a root, and then divide that factor out of the equation. This can be done as a straightforward division problem. Here is an example.

For $y = x^3 - 2x^2 - 2x - 3$, $y = 0$ at $x = 3$: $3^3 - (2)(3)^2 - (2)(3) - 3 = 0$. Therefore, $x - 3$ is a factor:

$$
\begin{array}{r}
x^2 \;\; + \;\; x \;\; + \;\; 1 \\
x - 3 \overline{)\, x^3 - 2x^2 - 2x - 3} \\
\underline{x^3 - 3x^2} \\
x^2 - 2x \\
\underline{x^2 - 3x} \\
x - 3 \\
\underline{x - 3} \\
0
\end{array}
$$

Knowing one root, then, can help you break the equation down and help you to find others. (*Note:* It obviously isn't necessary to factor out a root, since they all satisfy the original equation; however, it reduces the equation by 1 degree, thus making it much easier to find the next root.)

In case you feel bothered about dividing by zero in the example above ($x = 3$ is a root; therefore, $x - 3 = 0$), rest assured that you have not done so. The factor $x - 3$ is zero only at $x = 3$. Since we know that, it is no longer useful in finding other values of x for which the equation is zero. For every other value of x, it can be discarded. That is in fact just what we are doing. The other roots (there are exactly n roots in an nth-degree equation regardless of how many are identical) must make the other factor, in this case $x^2 + x + 1$, equal zero, but will not make $x - 3 = 0$. This is true even if there were two roots at $x = 3$ in the original equation. They both make the equation zero at $x = 3$ but cannot come out of the same factor $x - 3$ and must be factored out separately.

This procedure of factoring is mainly a matter of luck. With a little experience you can spot the obvious factors, but often an equation will prove unfactorable. Left with this possibility, you could try drawing a rough graphical sketch to see approximately where the roots are. Another recourse is to evaluate the function at various points and look for a change in sign. (If there's a change in sign, it must cross the x axis at *least* once.) Both of these methods have merit, but their usefulness wanes if great accuracy is desired simply because they become too clumsy. Furthermore, it is easy to miss a root or several roots if the various x values chosen are too widely spaced.

SAMPLES

Factor, and therefore find the roots of, the following:

1. $y = x^4 - 5x^2 + 4$ $(x + 1)(x - 1)(x + 2)(x - 2)$
2. $y = x^3 - 6x^2 + 11x - 6$ $(x - 1)(x - 2)(x - 3)$
3. $y = x^3 - 8x^2 - 3x + 90$ $(x + 3)(x - 5)(x - 6)$
4. $y = x^3 + 8x^2 + 8x + 7$ $(x + 7)(x^2 + x + 1)$

5. Sketch the curve to find to the nearest tenth the smallest positive and largest negative root of

$$y = x^3 - 5x^2 - 3x + 1 \qquad x = 0.2 \qquad x = -0.8$$

CHAPTER 7 SIMULTANEOUS EQUATIONS

Simultaneous equations is the name given to any set of equations that all have to be satisfied by the same set of values for the variables. This is a very broad definition, but in fact the major type consists of two or more variables in a linear combination: $a_{k1}x_1 + a_{k2}x_2 + \cdots + a_{kk}x_k = b$. In general, if there are n variables, it takes n and only n linearly *independent* equations of this form to obtain a unique solution, i.e., obtain only one value for each variable. Linearly independent means none of the equations can be made up by adding or subtracting some of the others to themselves or other equations any number of times.

$$x + y = 5 \quad 2x + 2y = 10 \qquad \text{are linearly } dependent$$

EXAMPLE

Solve for x, y.

$$x + y = 5$$
$$x - y = 1$$

There are two equations and two unknowns, so a solution is possible. To solve for x, add the two equations: $2x = 6$, $x = 3$. For y, subtract: $2y = 4$, $y = 2$, or just substitute the x value into one of the equations. It won't matter which one since the values must satisfy both: $3 + y = 5$, $y = 2$, or $3 - y = 1, y = 2$.

These simultaneous equations can be given graphical significance. For example the equations

$$3x + 2y = 7$$
$$x - 3y = 17$$

are both equations of straight lines. Although an infinite number of pairs x, y can satisfy one *or* the other, only one pair can satisfy both, the points where they intersect.

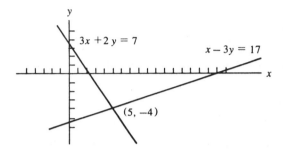

With three equations and three unknowns the point is the intersection of three planes. With more than three variables the graphical representations are cumbersome and harder to visualize, but analytically the solutions still have value. For instance, two spaceships can cross paths, but they can only rendezvous, or collide, if besides the three space coordinates the fourth variable, time, is also the same for both.

The easiest way to solve these types of simultaneous equations is to use the method employed in the first problem. What you will be doing is combining two of the equations into one by writing one of the variables in terms of the others. This is done by adding/subtracting a multiple of one equation to a multiple of the other.

$$2x - 3y = 7 \quad (1)$$
$$3x + 4y = 9 \quad (2)$$

$$3(1) \qquad 6x - 9y = 21$$
$$-2(2) \qquad \underline{-6x - 8y = -18}$$
$$0x - 17y = 3, \qquad y = -\frac{3}{17}$$

Plugging this value back into the equations or resolving them for x, you get $6x + \frac{27}{17} = 21$, $6x = 19\frac{7}{17}$, $x = 3\frac{4}{17}$. Naturally, with three equations and three unknowns you will have work to do. First, you will have to reduce the set into two equations, two unknowns by the above method, and then repeat the process to solve for one of the variables. Then go back to the intermediate set and solve for the second variable. Finally, go to the original set and get the value of the third variable. The more unknowns, the longer it takes and the more work involved.

Simultaneous equations can also be solved by a method using determinants. That method is described in the next section.

SIMULTANEOUS EQUATIONS

Solve for x, y.

1. $3x + 2y = 5$ $7x - 4y = 7$ $(\frac{17}{13}, \frac{7}{13})$
2. $x + 2y + 3z = 4$ $2x + 3y + z = 4$ $3x + y + 2z = 4$ $(\frac{2}{3}, \frac{2}{3}, \frac{2}{3})$
3. $5x - 4y = 6$ $2x - y = 1$ $(-\frac{2}{3}, -\frac{7}{3})$
4. If an object with mass, m_1, traveling with velocity, v_0, strikes an object at rest with mass, m_2, in a head-on elastic collision, both momentum and kinetic energy must be conserved. Each conservation law furnishes an equation:

$$m_1 v_0 = m_1 v_1 + m_2 v_2$$
$$\tfrac{1}{2} m_1 v_0^2 = \tfrac{1}{2} m_1 v_1^2 + \tfrac{1}{2} m_2 v_2^2$$

Solve for v_1 and v_2. Note the nature of the solutions for special cases: $m_1 = m_2$; $m_1 = \tfrac{1}{2} m_2$; $m_1 = 2m_2$; $m_1 \gg m_2$; $m_1 \ll m_2$. In particular note that since one of the equations is a quadratic, there must be two sets of solutions. For $m_1 = m_2$, one of these solutions is $v_1 = v_0$, $v_2 = 0$. The final velocity of the projectile is the same as its original velocity; the target continues at rest! Of course these values do conserve kinetic energy and momentum. Nowhere in the equations did we specify that there was a collision, and so the faithful mathematics includes such a possibility.

5. Set up the three simultaneous equations for the conservation of momentum and kinetic energy in two dimensions. Solve for the special case that $m_1 = m_2$ and show that $\theta_1 + \theta_2 = 90°$.

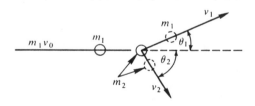

$$\Delta p_x = 0 \qquad m_1 v_0 = m_1 v_1 \cos \theta_1 + m_2 v_2 \cos \theta_2$$
$$\Delta p_y = 0 \qquad 0 = m_1 v_1 \sin \theta_1 - m_2 v_2 \sin \theta_2$$
$$\Delta E_{kin} = 0 \qquad \tfrac{1}{2} m_1 v_0^2 = \tfrac{1}{2} m_1 v_1^2 + \tfrac{1}{2} m_2 v_2^2$$

Since $m_1 = m_2$,

(1): $v_0 = v_1 \cos \theta_1 + v_2 \cos \theta_2$

(2): $0 = v_1 \sin \theta_1 - v_2 \sin \theta_2$

(3): $v_0^2 = v_1^2 + v_2^2$

$(1)^2$: $v_0^2 = v_1^2 \cos^2 \theta_1 + v_2^2 \cos^2 \theta_2 + 2v_1 v_2 \cos \theta_1 \cos \theta_2$

$(2)^2$: $0 = v_1^2 \sin^2 \theta_1 + v_2^2 \sin^2 \theta_2 - 2v_1 v_2 \sin \theta_1 \sin \theta_2$

$(1)^2 + (2)^2:$ $\quad v_0^2 = v_1^2 + v_2^2 + 2v_1 v_2(\cos \theta_1 \cos \theta_2 - \sin \theta_1 \sin \theta_2)$

(since $\sin^2 \theta + \cos^2 \theta = 1$)

Compare $(1)^2 + (2)^2$ with (3). It must be that $2v_1 v_2[\cos \theta_1 \cos \theta_2 - \sin \theta_1 \sin \theta_2] = 0$. Since in general neither v_1 nor v_2 is zero, it must be that $(\cos \theta_1 \cos \theta_2 - \sin \theta_1 \sin \theta_2) = 0$. A trig identity is that $(\cos \theta_1 \cos \theta_2 - \sin \theta_1 \sin \theta_2) = \cos(\theta_1 + \theta_2)$. But

$$\cos(\theta_1 + \theta_2) = 0 \quad \text{if } (\theta_1 + \theta_2) = 90°$$

A more elegant proof is geometrical. Consider the two vector momenta after the collision. Their vector sum must equal the original momentum.

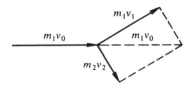

If $m_1 = m_2$, conservation of kinetic energy requires that $v_0^2 = v_1^2 + v_2^2$. That can be true only if the vector momenta form a right triangle:

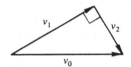 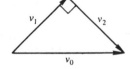 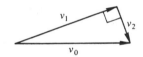

SIMULTANEOUS EQUATIONS

CHAPTER 8 DETERMINANTS

In the section on simultaneous equations we noted that there is a special way to solve these using determinants. Determinants were widely studied in the last half of the nineteenth century in connection with matrices and linear algebra. Theoretical interest in them is no longer what it was then, but some practical applications are still very important.

A. Matrices and determinants. A matrix is a rectangular array of numbers. Here are two ways of writing triple simultaneous equations:

$$a_{11}x_1 + a_{12}x_2 + a_{13}x_3 = b_1$$
$$a_{21}x_1 + a_{22}x_2 + a_{23}x_3 = b_2 \tag{8-1}$$
$$a_{31}x_1 + a_{32}x_2 + a_{33}x_3 = b_3$$

$$\begin{bmatrix} a_{11} & a_{12} & a_{13} \\ a_{21} & a_{22} & a_{23} \\ a_{31} & a_{32} & a_{33} \end{bmatrix} \begin{bmatrix} x_1 \\ x_2 \\ x_3 \end{bmatrix} = \begin{bmatrix} b_1 \\ b_2 \\ b_3 \end{bmatrix} \tag{8-2}$$

To multiply the array of constants and the vertical column of variables, let each top row element multiply a corresponding element of the variable in the vertical column; i.e.,

$$\begin{vmatrix} a_{11} & a_{12} & a_{13} \end{vmatrix} \begin{vmatrix} x_1 \\ x_2 \\ x_3 \end{vmatrix} \longrightarrow a_{11}x_1 + a_{12}x_2 + a_{13}x_3 = b_1$$

The next horizontal row is then used:

$$\begin{vmatrix} a_{21} & a_{22} & a_{23} \end{vmatrix} \begin{vmatrix} x_1 \\ x_2 \\ x_3 \end{vmatrix} \longrightarrow a_{21}x_1 + a_{22}x_2 + a_{23}x_3 = b_2$$

The array of constants is a matrix. Its determinant is a particular arrangement of products of those constants.

The determinant of a 1×1 matrix is equal to the matrix element. If the matrix is $[a]$ then $\det[a] = |a| = a$. The $|a|$ is just another way of writing determinant of a, and should not be confused with the absolute value of a, although the notation is the same. The determinant of a 2×2 matrix $M = \begin{bmatrix} a & c \\ b & d \end{bmatrix}$ is $\begin{vmatrix} a & c \\ b & d \end{vmatrix} = ad - bc$. This is not as mysterious as it looks. The determinant can be evaluated using a diagonal rule. You form products of entries on a diagonal. For diagonals slanting *down* to the right, the result is the product. In matrix M there is only one: ad. For diagonals slanting *up* to the right the result is the negative of the product: $-(bc)$. The determinant equals the sum of all these products:

$$\begin{vmatrix} a & c \\ b & d \end{vmatrix} = ad - bc \tag{8-3}$$

The diagonal rule holds for a 3×3 matrix also, but now there have to be three terms in a product.

$$\begin{vmatrix} a & b & c \\ d & e & f \\ g & h & i \end{vmatrix} = aei + bfg + cdh - gec - hfa - idb \tag{8-4}$$

Perhaps this can be visualized more easily if two of the rows are added on again and only full diagonals are included:

What is the determinant of $\begin{vmatrix} 3 & 5 \\ 7 & 9 \end{vmatrix}$?   *Ans.* -8.

What is the determinant of $\begin{vmatrix} 1 & 2 & 3 \\ 4 & 5 & 6 \\ 7 & 8 & 9 \end{vmatrix}$?   *Ans.* 0.

The diagonal rule gives $97 + 84 + 45 = 225$ and $-48 - 72 - 105 = -225$.

SAMPLES

Evaluate

DETERMINANTS

1. $\begin{vmatrix} 2 & 4 \\ 1 & 5 \end{vmatrix}$ (6.)

2. $\begin{vmatrix} 1 & 2 & 3 \\ 1 & 3 & 5 \\ 1 & 5 & 6 \end{vmatrix}$ (−3.)

3. $\begin{vmatrix} 7 & 3 \\ 2 & 6 \end{vmatrix}$ (36.)

4. $\begin{vmatrix} 13 & 6 \\ 11 & 9 \end{vmatrix}$ (51.)

5. $\begin{vmatrix} 4 & 11 & 2 \\ 5 & 10 & 9 \\ 16 & 6 & 8 \end{vmatrix}$ (988.)

B. Cofactors of determinants. Another fact about determinants is that the diagonal rule only applies to determinants of order 3 or less. For higher-order determinants another method can be used. We shall apply it to a third-order determinant to illustrate it, but it will work for a determinant of any order. First it may be more convenient to write our determinants this way:

$$\begin{vmatrix} a_{11} & a_{12} & a_{13} \\ a_{21} & a_{22} & a_{23} \\ a_{31} & a_{32} & a_{33} \end{vmatrix}$$

where the first subscript identifies the row and the second identifies the column that the element is in. The determinant of an $n \times n$ matrix A can be written as

$$\det A = \sum_{i=1}^{n} (-1)^{i+j}(a_{ij})A_{ij} \qquad (8\text{--}5)$$

where $(-1)^{i+j}A_{ij}$ is called a *cofactor* or signed minor of the determinant.

The cofactor A_{ij} is just the original determinant without the ith row and jth column.

$$\begin{vmatrix} a_{11} & a_{12} & \cdots & a_{1j-1} & a_{1j} & a_{1j+1} & \cdots & a_{1n} \\ a_{21} & \cdot & \cdots & \cdot & a_{2j} & \cdot & \cdots & a_{2n} \\ \cdot & \cdot & \cdots & \cdot & \cdot & \cdot & \cdots & \cdot \\ \cdot & \cdot & \cdots & \cdot & \cdot & \cdot & \cdots & \cdot \\ a_{i-1,1} & & \cdots & \cdot & \cdot & \cdot & \cdots & \cdot \\ a_{i1} & a_{i2} & \cdots & \cdot & a_{ij} & \cdot & \cdots & \cdot \\ a_{i+1,1} & & \cdots & \cdot & \cdot & \cdot & \cdots & \cdot \\ \cdot & \cdot & \cdots & \cdot & \cdot & \cdot & \cdots & \cdot \\ \cdot & \cdot & \cdots & \cdot & \cdot & \cdot & \cdots & \cdot \\ a_{n1} & \cdot & \cdots & \cdot & \cdot & \cdot & \cdots & a_{nn} \end{vmatrix}$$

DETERMINANTS

$$\longrightarrow (-1)^{i+j} \begin{vmatrix} a_{11} & \cdots & a_{1j-1} & a_{1j+1} & \cdots & a_{1n} \\ \cdot & \cdots & \cdot & \cdot & \cdots & \cdot \\ \cdot & \cdots & \cdot & \cdot & \cdots & \cdot \\ \cdot & \cdots & \cdot & \cdot & \cdots & \cdot \\ a_{i-1,1} & \cdots & \cdot & \cdot & \cdots & \cdot \\ a_{i+1,1} & \cdots & \cdot & \cdot & \cdots & \cdot \\ \cdot & \cdots & \cdot & \cdot & \cdots & \cdot \\ \cdot & \cdots & \cdot & \cdot & \cdots & \cdot \\ \cdot & \cdots & \cdot & \cdot & \cdots & \cdot \\ a_{n1} & \cdots & \cdot & \cdot & \cdots & a_{nn} \end{vmatrix}$$

The cofactor A_{ij} is therefore a determinant of order $n-1$ since it has one less row and one less column. The summation (sigma) says to do this down a column, but the determinant is also equal to $\det A = \sum_{j=1}^{n} (a_{ij})(-1)^{i+j}A_{ij}$. In other words, you can pick any column or row and evaluate the various sums along the particular column or row. [Often the term $(-1)^{i+j}A_{ij}$ is written as just A_{ij}, where it is assumed the sign is incorporated into the minor A_{ij}.]

EXAMPLE

Earlier, using the diagonal rule, we found $\begin{vmatrix} 1 & 2 & 3 \\ 4 & 5 & 6 \\ 7 & 8 & 9 \end{vmatrix} = 0.$

We can also do this with cofactors. Doing it along the first row we have

$(1)(-1)^{1+1}\begin{vmatrix} 5 & 6 \\ 8 & 9 \end{vmatrix} + 2(-1)^{1+2}\begin{vmatrix} 4 & 6 \\ 7 & 9 \end{vmatrix} + 3(-1)^{1+3}\begin{vmatrix} 4 & 5 \\ 7 & 8 \end{vmatrix} = 1(1)(45 - 48)$

$+ 2(-1)(36 - 42) + 3(1)(32 - 35) = -3 + 12 - 9 = 0$. The work may seem easier because the determinants are of order 2 and with practice can be done in your head.

There are shortcuts and tricks for evaluating determinants. Anything said in this section about rows applies equally to columns.

If any determinant has an entire row of zeros, then the determinant is zero because we can expand the determinant along that row, and no matter what the cofactors equal, each is multiplied by zero so that the determinant is zero. This gives a hint as to how to expand a determinant, for if a row is mostly filled with zeros there will be only a few terms, the rest being zero.

C. Cramer's rule for solving simultaneous equations. A determinant will equal zero if one of its rows or columns is a linear combination of the other rows or columns, respectively—in other words, if it is linearly dependent. This idea of linear dependence sounds like the condition for simultaneous

DETERMINANTS

equations. Determinants can indeed be used to solve simultaneous equations using Cramer's rule. The rule is

$$x_i = \frac{\begin{array}{ccccccc} a_{11} & \cdots & a_{1i-1} & b_1 & a_{1i+1} & \cdots & a_{1n} \\ \cdot & & & & & & \cdot \\ \cdot & & & & & & \cdot \\ \cdot & & & & & & \cdot \\ a_{n1} & \cdots & a_{ni-1} & b_n & a_{ni+1} & \cdots & a_{nn} \end{array}}{\begin{array}{ccccccc} a_{11} & \cdot & \cdot & \cdot & \cdot & \cdot & a_{1n} \\ \cdot & & & & & & \cdot \\ \cdot & & & & & & \cdot \\ \cdot & & & & & & \cdot \\ a_{n1} & \cdot & \cdot & \cdot & \cdot & \cdot & a_{nn} \end{array}} \tag{8-6}$$

For triple equations the rule gives these solutions:

$$x_1 = \frac{\begin{vmatrix} b_1 & a_{12} & a_{13} \\ b_2 & a_{22} & a_{23} \\ b_3 & a_{32} & a_{33} \end{vmatrix}}{\begin{vmatrix} a_{11} & a_{12} & a_{13} \\ a_{21} & a_{22} & a_{23} \\ a_{31} & a_{32} & a_{33} \end{vmatrix}} \qquad x_2 = \frac{\begin{vmatrix} a_{11} & b_1 & a_{13} \\ a_{21} & b_2 & a_{23} \\ a_{31} & b_3 & a_{33} \end{vmatrix}}{\begin{vmatrix} a_{11} & a_{12} & a_{13} \\ a_{21} & a_{22} & a_{23} \\ a_{31} & a_{32} & a_{33} \end{vmatrix}}$$

$$x_3 = \frac{\begin{vmatrix} a_{11} & a_{12} & b_1 \\ a_{21} & a_{22} & b_2 \\ a_{31} & a_{32} & b_3 \end{vmatrix}}{\begin{vmatrix} a_{11} & a_{12} & a_{13} \\ a_{21} & a_{22} & a_{23} \\ a_{31} & a_{32} & a_{33} \end{vmatrix}} \tag{8-7}$$

It says to replace the coefficients of the variable you are solving for with the corresponding constants on the right side of the equation, and divide this determinant by the determinant of the coefficients. If the equations are linearly independent, the determinant in the denominator will not be zero and you will be able to solve for all the variables. This method is useful if you are only interested in one or a few variables, but it is tedious even for a high-speed computer to do it for very many variables. Notice that the denominator is the same for all the variables and only needs to be evaluated once.

EXAMPLE

Solve for x, y using Cramer's rule.

$$2x + 3y = 5$$
$$7x - 4y = 13$$

$$x = \frac{\begin{vmatrix} 5 & 3 \\ 13 & -4 \end{vmatrix}}{\begin{vmatrix} 2 & 3 \\ 7 & -4 \end{vmatrix}} = \frac{-59}{-29} \qquad y = \frac{\begin{vmatrix} 2 & 5 \\ 7 & 13 \end{vmatrix}}{-29} = \frac{-9}{-29}$$

Solve $ax + by = c$, $dx + ey = f$, and get a general solution of the two-variable problem using Cramer's rule.

$$x = \frac{\begin{vmatrix} c & b \\ f & e \end{vmatrix}}{\begin{vmatrix} a & b \\ d & e \end{vmatrix}} = \frac{ce - bf}{ae - bd} \qquad y = \frac{\begin{vmatrix} a & c \\ d & f \end{vmatrix}}{ae - bd} = \frac{af - cd}{ae - bd}$$

There are other applications of determinants, such as the representation of areas and perhaps most importantly the cross product of vectors. We shall defer discussion of these until a later chapter, however. (See Section 10.2.)

SAMPLES

Use the method of cofactors to evaluate the following determinants:

1. $\begin{vmatrix} 6 & 3 & 7 \\ 0 & 3 & 5 \\ 4 & 9 & 8 \end{vmatrix}$ $[6(-21) + 4(-6) = -150.]$

2. $\begin{vmatrix} 2 & 8 & 7 & 5 \\ 9 & 1 & 6 & 6 \\ 5 & 2 & 0 & 5 \\ 3 & 1 & 4 & 2 \end{vmatrix}$

$[7\begin{vmatrix} 9 & 1 & 6 \\ 5 & 2 & 5 \\ 3 & 1 & 2 \end{vmatrix} - 6\begin{vmatrix} 2 & 8 & 5 \\ 5 & 2 & 5 \\ 3 & 1 & 2 \end{vmatrix} - 4\begin{vmatrix} 2 & 8 & 5 \\ 9 & 1 & 6 \\ 5 & 2 & 5 \end{vmatrix}$

$= 7(81 - 91) - 6(153 - 120) - 4(340 - 409)$
$= -70 - 198 + 276 = 8.]$

3. $\begin{vmatrix} 2 & 0 & 2 & 6 \\ 1 & 4 & 9 & 2 \\ 3 & 7 & 6 & 5 \\ 8 & 1 & 9 & 2 \end{vmatrix}$

$[2\begin{vmatrix} 4 & 9 & 2 \\ 7 & 6 & 5 \\ 1 & 9 & 2 \end{vmatrix} + 2\begin{vmatrix} 1 & 4 & 2 \\ 3 & 7 & 5 \\ 8 & 1 & 2 \end{vmatrix} - 6\begin{vmatrix} 1 & 4 & 9 \\ 3 & 7 & 6 \\ 8 & 1 & 9 \end{vmatrix}$

$= 2(219 - 318) + 2(180 - 141) - 6(282 - 618)$
$= -198 + 78 + 2016 = 1896.]$

Use Cramer's rule to solve the following systems of simultaneous equations:

4. $2x - 3y + 2z = -1$
     $-x + 7y - z = 17$
     $5x + 2y + 3z = 4$ $(x = -7, y = 3, z = 11.)$

5. $2x - y + z - t = -1$
     $4x + 4y - 6z + 3t = 35$
     $y + z + t = 4$
     $-2x + 3y + 4z - 3t = 0$ $(x = 2.5, y = 4, z = -1, t = 1.)$

CHAPTER9 GEOMETRY

9.1 ANALYTICAL GEOMETRY. One of the greatest advances in mathematics came during the sixteenth century when René Descartes began drawing geometrical figures on graphs. Geometry and algebra were linked. With a simple device of a grid of lines, it is possible to locate any point merely by specifying its distance and direction from a common point called the origin.

1. With Cartesian coordinates in two dimensions, you locate points by constructing a horizontal reference direction, called the x axis, and a vertical reference direction, called the y axis, and listing the location of any point in terms of distances along each axis: (x, y). Notice that the point $(1, 2)$ is not

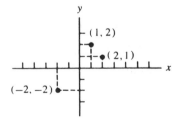

the same as $(2,1)$. The x distance, the abscissa, is always listed first, then the y distance, the ordinate. In the case of three dimensions, the coordinates are listed (x, y, z) and the graph looks like the corner of a room where two walls and a floor meet.

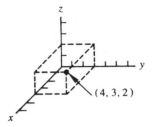

(4, 3, 2)

2. The Cartesian axes are mutually perpendicular, which means that the Pythagorean theorem for right triangles holds. Therefore, in two dimensions the distance between two points (x_1, y_1) and (x_2, y_2) is

$$d = \sqrt{(x_2 - x_1)^2 + (y_2 - y_1)^2} \tag{9-1}$$

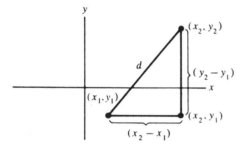

In three dimensions the analogous expression is

$$d = \sqrt{(x_2 - x_1)^2 + (y_2 - y_1)^2 + (z_2 - z_1)^2} \tag{9-2}$$

and so on for 4, 5, . . . , n dimensions, although the geometric significance is then hard to visualize.

3. An important property of any curve drawn on a graph is its slope. For a straight line, $y = mx + b$, the slope is m, defined as "the rise over the run" or "the change in y over the change in x" or $(y_2 - y_1)/(x_2 - x_1)$ or $\Delta y / \Delta x$. With curves other than straight lines the slope retains this general meaning with the qualification that the slope is taken as the limiting value as x_2 approaches x_1. If such a limit exists, it is called the derivative of the curve at the point x_1.

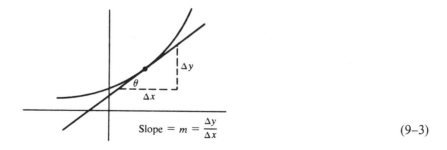

$$\text{Slope} = m = \frac{\Delta y}{\Delta x} \tag{9-3}$$

123

4. From the graph you can see that Δy and Δx form the legs of a right triangle, and the ratio defined by the slope $\Delta y/\Delta x$ is also $\tan \theta$. If two lines intersect, the tangent of the angle between them is $\tan \theta = \tan(\theta_2 - \theta_1) = (\tan \theta_2 - \tan \theta_1)/(1 + \tan \theta_1 \tan \theta_2)$, or since the slopes equal the tangents,

$$\tan \theta = \frac{m_2 - m_1}{1 + m_1 m_2} \tag{9-4}$$

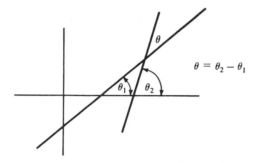

$$\theta = \theta_2 - \theta_1$$

If two lines are perpendicular, the angle between them is 90°, and $\tan \theta = \infty$. Therefore, the denominator of the last expression would have to equal zero (unless $m_1 = 0$, in which case the second line would be vertical with $m_2 = \infty$). For the denominator to equal zero, and therefore for any two lines to be perpendicular:

$$m_1 m_2 = -1 \quad \text{(for perpendicular lines)} \tag{9-5}$$

Another way to prove this is to make use of the following diagram:

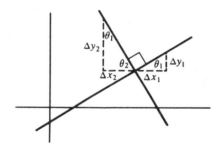

$$\tan \theta_2 = \cot \theta_1$$
$$\tan \theta_1 \tan \theta_2 = \tan \theta_1 \cot \theta_1 = -1$$

(There is a minus sign because if one slope is positive, the other must be negative.)

5. The equations of lines tangent and perpendicular to a curve at a given point are found in the following manner. Since the slope of a line at a point is given by dy/dx evaluated at that point, the slope of the perpendicular line at that point must be $-[1/(dy/dx)]$. Hence the equation of the tangent line through (x_1, y_1) is

$$y - y_1 = \left(\frac{dy}{dx}\right)_{x_1, y_1} (x - x_1) \tag{9-6}$$

and the equation of the normal (perpendicular) line is

$$y - y_1 = -\frac{1}{(dy/dx)_{x_1 y_1}}(x - x_1) \tag{9-7}$$

For example, at $(x, y) = (2, 2)$ on the parabola, $y^2 = 2x$, $(dy/dx)_{2,2} = (1/y)_{2,2} = \frac{1}{2}$. The equation of the tangent line is

$$y - 2 = \tfrac{1}{2}(x - 2) \qquad \text{or} \qquad y = \tfrac{1}{2}x + 1$$

and the equation of the normal at this point is

$$(y - 2) = -2(x - 2) \qquad \text{or} \qquad y = -2x + 6.$$

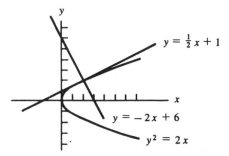

6. The description of a line in three dimensions can be given in terms of the coordinates of any two points through which the line passes: $(x_1 y_1 z_1)$ and $(x_2 y_2 z_2)$. The line can also be specified by describing its direction and giving the coordinates of one point. The direction is usually given with respect to the Cartesian axes in terms of the direction cosines, rather than the angles themselves. Notice that these angles are between the line and the axes, and are not the spherical coordinate angles (except that $\gamma = \theta$).

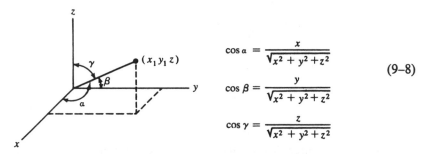

$$\cos \alpha = \frac{x}{\sqrt{x^2 + y^2 + z^2}}$$

$$\cos \beta = \frac{y}{\sqrt{x^2 + y^2 + z^2}} \tag{9-8}$$

$$\cos \gamma = \frac{z}{\sqrt{x^2 + y^2 + z^2}}$$

These direction cosines apply to any line that is parallel to this line. For some line that does not pass through the origin,

$$\cos \alpha = \frac{x_2 - x_1}{d}$$

$$\cos \beta = \frac{y_2 - y_1}{d} \tag{9-9}$$

$$\cos \gamma = \frac{z_2 - z_1}{d}$$

Any three numbers proportional to these direction cosines also define the same direction:

$$\cos \alpha : \cos \beta : \cos \gamma = a : b : c$$

$$\cos \alpha = \frac{a}{\sqrt{a^2 + b^2 + c^2}} \qquad \cos \beta = \frac{b}{\sqrt{a^2 + b^2 + c^2}}$$

$$\cos \gamma = \frac{c}{\sqrt{a^2 + b^2 + c^2}} \qquad (9\text{--}10)$$

This relationship holds, since

$$\cos^2 \alpha + \cos^2 \beta + \cos^2 \gamma = \frac{(x_2 - x_1)^2}{d^2} + \frac{(y_2 - y_1)^2}{d^2}$$

$$+ \frac{(z_2 - z_1)^2}{d^2} = 1 \qquad (9\text{--}11)$$

The line can also be defined by

$$x - x_1 = k(y - y_1) = l(z - z_1) \qquad (9\text{--}12)$$

7. The angle between two lines is given by

$$\cos \theta = \cos \alpha_1 \cos \alpha_2 + \cos \beta_1 \cos \beta_2 + \cos \gamma_1 \cos \gamma_2 \qquad (9\text{--}13)$$

This angle, θ, is the angle between two lines parallel to the given ones and passing through the origin.

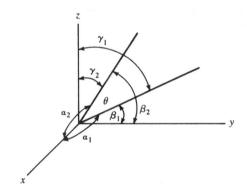

A vector in three-dimensional space can be described as

$$\mathbf{A}_1 = |A_1| [\cos \alpha_1 \hat{i} + \cos \beta_1 \hat{j} + \cos \gamma_1 \hat{k}] \qquad (9\text{--}14)$$

The dot product of two vectors is

$$\mathbf{A}_1 \cdot \mathbf{A}_2 = |A_1| |A_2| \cos \theta \qquad (9\text{--}15)$$

where θ is the angle between $\mathbf{A}_1$ and $\mathbf{A}_2$. Consequently,

$$\mathbf{A}_1 \cdot \mathbf{A}_2 = |A_1| |A_2| \cos \theta = |A_1| |A_2| [\cos \alpha_1 \cos \alpha_2$$

$$+ \cos \beta_1 \cos \beta_2 + \cos \gamma_1 \cos \gamma_2] \qquad (9\text{--}16)$$

This demonstrates in another way that

$$\cos \theta = \cos \alpha_1 \cos \alpha_2 + \cos \beta_1 \cos \beta_2 + \cos \gamma_1 \cos \gamma_2$$

If a, b, c and d, e, f are trios of numbers proportional to the direction cosines of the two lines, and if $ad + be + cf = 0$, then $\cos \theta = 0$, and $\theta = 90°$. The lines are perpendicular to each other.

EXAMPLE

What is the angle between the line passing through the origin and the point $(5, 4, 3)$, and the line passing through $(2, 3, 9)$ and $(4, 5, 3)$?

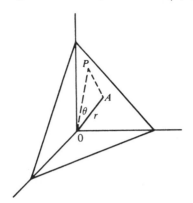

$$\cos a_1 = \frac{5}{\sqrt{25 + 16 + 9}} = \frac{5}{\sqrt{50}} , \cos \beta_1 = \frac{4}{\sqrt{50}} , \cos \gamma_1 = \frac{3}{\sqrt{50}}$$

$$\cos a_2 = \frac{4 - 2}{\sqrt{2^2 + 2^2 + (-6)^2}} = \frac{2}{\sqrt{44}} , \cos \beta_2 = \frac{2}{\sqrt{44}} , \cos \gamma_2 = \frac{-6}{\sqrt{44}}$$

$$\cos \theta = \frac{5}{\sqrt{50}} \frac{2}{\sqrt{44}} + \frac{4}{\sqrt{50}} \frac{2}{\sqrt{44}} + \frac{3}{\sqrt{50}} \frac{(-6)}{\sqrt{44}} = 0$$

$$\theta = 90°$$

Any plane can be described by specifying the direction cosines of a line perpendicular to the plane going through the origin, and giving the distance of the plane from the origin. The perpendicular to the plane intersects it at A and has direction cosines α, β, γ. The direction cosines of a line to any other point on the plane are $\alpha_1, \beta_1, \gamma_1$.

$$OA = r = OP \cos \theta = OP (\cos \alpha_1 \cos \alpha + \cos \beta_1 \cos \beta$$
$$+ \cos \gamma_1 \cos \gamma) \qquad (9\text{–}17)$$

Since $\cos \alpha_1 = x/OP$, $\cos \beta_1 = y/OP$, and $\cos \gamma_1 = z/OP$,

$$r = x \cos \alpha + y \cos \beta + z \cos \gamma \qquad (9\text{–}18)$$

9.1 ANALYTICAL GEOMETRY

Any equation of this general form describes a plane:

$$Ax + By + Cz + D = 0 \qquad (9\text{--}19)$$

The direction cosines of the normal line are

$$\cos \alpha = \frac{A}{\sqrt{A^2 + B^2 + C^2}} \qquad \cos \beta = \frac{B}{\sqrt{A^2 + B^2 + C^2}}$$

$$\cos \gamma = \frac{C}{\sqrt{A^2 + B^2 + C^2}} \qquad (9\text{--}20)$$

In vector form the normal line has the form

$$\mathbf{N} = A\hat{i} + B\hat{j} + C\hat{k} \qquad (9\text{--}21)$$

The distance from the origin to the plane is

$$r = -\frac{D}{\sqrt{A^2 + B^2 + C^2}} \qquad (9\text{--}22)$$

EXAMPLES

1. What is the equation of the plane cutting the axes symmetrically at the points (3, 0, 0), (0, 3, 0), and (0, 0, 3)?

$$3A + 0 + 0 = -D$$
$$0 + 3B + 0 = -D \qquad A = B = C = -\frac{D}{3}$$
$$0 + 0 + 3C = -D$$

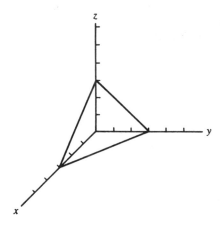

The desired equation is $x + y + z = 3$. The direction cosines of the normal to the plane are $1/\sqrt{3} : 1/\sqrt{3} : 1/\sqrt{3}$, and the length of the normal is $3/\sqrt{3} = \sqrt{3}$.

2. The equation of a plane parallel to the xy plane, and at a distance k from it, is

$$z = k$$

Since the normal line is perpendicular to the x axis and the y axis, $\cos \alpha = \cos \beta = 0$.

3. The equation of plane through the z axis and bisecting the xy axes is

$$x - y = 0$$

Since the normal line is perpendicular to the z axis, $\cos \gamma = 0$, and at every point on the bisecting plane, $x = y$.

9.2 CONIC SECTIONS. A great many geometrical figures describe the motion or shape of physical objects. Let us examine one class of these called conic sections. Conic sections are plane figures obtained by cutting a right circular cone with a plane. Cut the cone perpendicular to the axis and the

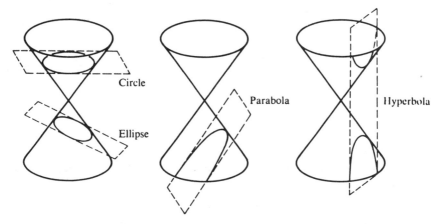

Circle

Ellipse

Parabola

Hyperbola

intersection is a circle of arbitrary size, depending on where the cone was cut. If the cutting plane is tipped at an angle, an out-of-round circle appears, called an ellipse. At a greater angle, parallel to a side of the cone, the closed curve breaks open and becomes a parabola. At a still greater angle the curve becomes one side of a hyperbola.

1. The expression for a *circle* centered at the origin is

$$x^2 + y^2 = R^2 \tag{9-23}$$

where R is the radius. If the center is at some point (a, b), the expression becomes

$$(x - a)^2 + (y - b)^2 = R^2 \tag{9-24}$$

By completing the squares we can turn this equation into the general form

$$x^2 + y^2 + Cx + Dy + E = 0 \tag{9-25}$$

Since there are three constants that determine the equation, three conditions can be satisfied about the size and location of any circle. In particular, we can find a circle to go through any three arbitrary points (not in a straight line). The equation with its three undetermined constants must satisfy each of the three points:

$$x_1^2 + y_1^2 + Cx_1 + Dy_1 + E = 0$$
$$x_2^2 + y_2^2 + Cx_2 + Dy_2 + E = 0 \qquad (9\text{–}26)$$
$$x_3^2 + y_3^2 + Cx_3 + Dy_3 + E = 0$$

Note that these three simultaneous equations are to be solved for C, D, and E. The (x_i, y_i) are given and become constants.

Another way to describe a circle is by means of a parametric set of equations. Each of the coordinates (x and y) is described in terms of a third variable, a parameter.

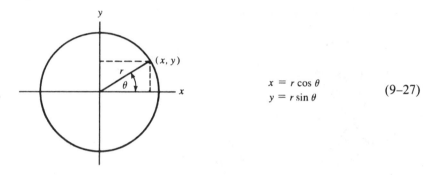

$$x = r \cos \theta$$
$$y = r \sin \theta \qquad (9\text{–}27)$$

In this case the parameter, θ, can be eliminated by squaring and adding the two equations.

$$x^2 + y^2 = r^2(\cos^2 \theta + \sin^2 \theta) = r^2 \qquad (9\text{–}28)$$

2. An *ellipse* is the locus of points whose sum of distances from the foci is a constant. In the diagram we have placed the two foci on the x axis. The constant distance is assigned the value $2a$, where $a > c$.

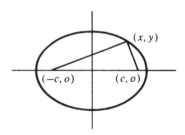

Adding up the two distances from the foci to (x, y),

$$\sqrt{(x + c)^2 + y^2} + \sqrt{(c - x)^2 + y^2} = 2a$$
$$(x + c)^2 + y^2 = [2a - \sqrt{(c - x)^2 + y^2}]^2$$
$$= 4a^2 + (c - x)^2 + y^2$$
$$- 4a\sqrt{(c - x)^2 + y^2}$$

$$xc - a^2 = -a\sqrt{(c - x)^2 + y^2}$$
$$x^2c^2 + a^4 - 2xca^2 = a^2c^2 + a^2x^2 - 2cxa^2 + a^2y^2$$
$$x^2(a^2 - c^2) + y^2a^2 = a^2(a^2 - c^2)$$

Let $b^2 = (a^2 - c^2)$, and divide by a^2b^2:

$$\frac{x^2}{a^2} + \frac{y^2}{b^2} = 1 \tag{9-29}$$

For an ellipse centered at (e, f), the equation is

$$\frac{(x - e)^2}{a^2} + \frac{(y - f)^2}{b^2} = 1 \tag{9-30}$$

The general form of the equation is

$$Ax^2 + By^2 + Cx + Dy + E = 0 \qquad \text{where } A \text{ and } B$$
$$\text{are positive} \tag{9-31}$$

To analyze the curve, solve the origin-centered equation for y:

$$y = \pm \frac{b}{a}\sqrt{a^2 - x^2} \tag{9-32}$$

There are real values for y only when $|x| \le a$. When $x = a$, $y = 0$. When $x = 0$, $y = \pm b$. The curve is not symmetric around a point like a circle, but is symmetric around the major and minor axes. The length of the major axis is $2a$, and the length of the minor axis is $2b$. The foci are located at $\pm c$ on the major axis. You can see that the sum of the distances from the foci to a point on the curve equals $2a$ by considering the distances to the extreme right or left points. From this, by way of the Pythagorean theorem, $b^2 = a^2 - c^2$. A convenient way to draw an ellipse is to fasten thumbtacks or nails to the two focal points and tie a string between them with a length of $2a$. Place a pencil point in the loop of string and draw the curve that always keeps the string taut. The sum of the distances from the foci to any point on the curve is automatically kept constant at $2a$, since that is the length of the string.

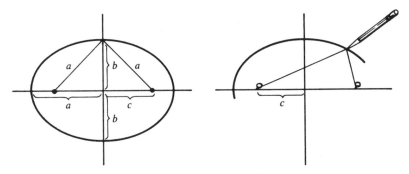

The parametric equations for an ellipse centered at the origin, with major and minor axes a and b, are

$$x = a \cos \theta \qquad y = b \sin \theta \qquad\qquad (9\text{--}33)$$

The ellipse lies between two circles with radii a and b:

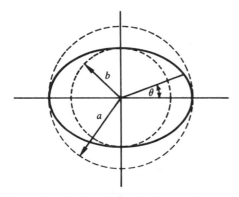

The ratio c/a is called the eccentricity, e, of the ellipse and is a measure of the out-of-roundness. For an ellipse, $0 \le e \le 1$. In the limiting case of a circle, $c = 0$, $b = a$, and $e = 0$. The eccentricity of the earth's solar orbit is 0.0167. Mercury has a very large eccentricity of 0.2056. The eccentricity of the moon's orbit around the earth is 0.055. Note the significance of these values of eccentricity for purposes of visualizing or modeling the orbits. For the earth's orbit, $e = 0.0167 = c/a$, and $b^2 = a^2 - c^2 = a^2(1 - e^2) = a^2(1 - 2.79 \times 10^{-4})$. Since b^2/a^2 differs from 1 by less than 3 parts per 10^4, b/a differs by less than $1\frac{1}{2}$ parts per 10^4. Evidently, the earth's orbit appears circular. The focus is at a distance of $c = ea = 0.0167 \times 93 \times 10^6$ miles $= 1.6 \times 10^6$ miles from the center. Since the diameter of the sun is about 10^6 miles, the center of the sun, which is at one of the foci of the earth's orbit, is very close to the center of the ellipse. Ellipses with three different eccentricities are shown in the diagram.

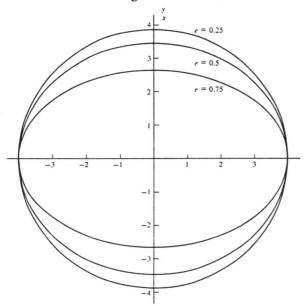

If wave motion originates at one focus of an ellipse, it will focus on the other. This is required by the properties of the ellipse and the nature of focusing. The distance from one focus to any point on the surface and then back to the other focus is a constant, by definition of the ellipse. Hence segments of a pulse starting out together from one focus reach the walls at varying times but can all arrive at the other focus at the same time, and so in phase. Focusing geometries can be analyzed either in terms of detailed tracing of angles of reflection and refraction, or by seeking paths of equal time of passage. Three-dimensional examples of elliptical (ellipsoidal) focusing are well known in large galleries or "whispering" domes. There is one in the capitol building in Washington, and another in the Mormon Tabernacle in Salt Lake City.

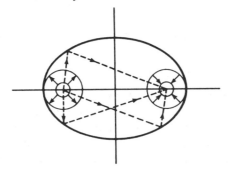

3. A *parabola* is defined as the locus of points equidistant from a focus and a line called the directrix. (The conic sections can also be expressed in polar coordinates with a different definition; the definition of the parabola, however, remains essentially the same.)

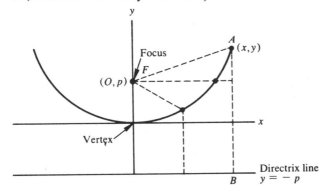

From the construction, these conditions lead to the following equation

$$FA = AB$$
$$\sqrt{x^2 + (y - p)^2} = y + p$$
$$x^2 + y^2 - 2py + p^2 = y^2 + p^2 + 2py \tag{9-34}$$
$$x^2 = 4py$$

The formula above is for a y-oriented parabola with the vertex at the origin. Since the vertex is equidistant from the focus and the directrix, the focus is at (O, p) and the directrix is at $y = -p$. For the vertex of the parabola to be at a point (a, b), the formula becomes $(x - a)^2 = 4p(y - b)$. The focus would then be at $(a, b + p)$ and the directrix at $y = b - p$.

The general form for a parabola symmetric about a line parallel to the y axis is

$$Ax^2 + Cx + Dy + E = 0 \qquad \text{where } A \text{ is positive} \qquad (9\text{--}35)$$

Similarly, for a parabola symmetric about a line parallel to the x axis,

$$By^2 + Cx + Dy + E = 0 \qquad \text{where } B \text{ is positive} \qquad (9\text{--}36)$$

Parametric equations for a parabola are commonly used in the description of trajectories. Horizontal and vertical components of velocity are independent of each other. Neglecting air friction, earth spin, etc., the horizontal and vertical displacements of a projectile are given by

$$\Delta x = v_x \, \Delta t \qquad \Delta y = \tfrac{1}{2}g(\Delta t)^2 \quad \text{where } (v_y)_0 = 0$$

Eliminating Δt, the time interval that serves as a parameter, $\Delta y = \tfrac{1}{2}g(\Delta x / v_x)^2$. This is the standard form for a parabolic relationship between Δx and Δy.

The parabola is usually a first approximation to the shape of a potential energy well. This is because the restoring force for any small distortion of a bound system is usually proportional to the amount of the distortion. If $F \propto \Delta x$, then $E_{pot} \propto \Delta x^2$.

For a simple suspension bridge with constant loading along the horizontal direction, the cable hangs in the form of a parabola.

As we demonstrated with the parametric equations, the short-range trajectories of projectiles are parabolic. The requirements are that air resistance and rotational forces be negligible so that horizontal displacement is proportional to the first power of time and vertical displacement depends on the square of the time. The range must be short, since the conditions assume that the earth is flat and gravity acts vertically.

Reflecting telescope mirrors are parabolic. The reasoning is parallel to that used in the case of the ellipse, but now we make use of the different condition of parabolic geometry.

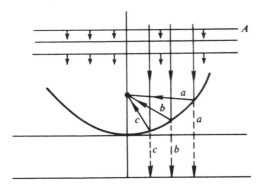

Parallel rays representing wave fronts (lines of constant phase) come in parallel to the axis. All sections of wave front A would reach the directrix at the same time if no mirror were in the way. Because of the definition of the parabola, the distance from the mirror at any point to the directrix is the same as the distance from that point to the focus. Thus the wave front can be reflected so that all of it arrives in phase at the parabolic focus, making it also the wave focus. In practice, large telescope mirrors are first ground to a spherical shape and then deepened slightly to form a paraboloid.

4. The *hyperbola* is defined as the locus of points whose *difference* of distances from two foci is a constant.

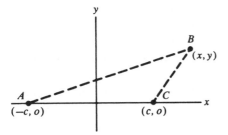

Imposing this condition on our standard construction leads to this equation:

$$AB - CB = 2a$$

$$\sqrt{(x + c)^2 + y^2} - \sqrt{(x - c)^2 + y^2} = 2a$$

Following the same process of squaring and reducing that we used in the case of the ellipse, we get

$$\frac{x^2}{a^2} - \frac{y^2}{b^2} = 1 \qquad \text{where } b^2 = c^2 - a^2 \tag{9-37}$$

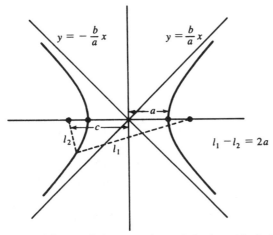

The general form of the equation of the hyperbola is

$$Ax^2 - By^2 + Cx + Dy + E = 0 \qquad \text{where } A \text{ and } B \text{ are positive} \tag{9-38}$$

The formula is similar to that of the ellipse, the difference consisting of a minus sign instead of a plus sign in front of the y^2 term. Again $2a$ is the length of the major axis (from vertex to vertex) and is also the constant difference in distances between any point on the curve and the two foci, as can be seen by examining the conditions at either vertex in the diagram.

The two curves in the figure are mirror images, both satisfying the imposed conditions. Solving for y, we have

$$y = \pm \frac{b}{a}\sqrt{x^2 - a^2} \tag{9-39}$$

For real values of y, x is restricted to values equal to or greater than a. $|x| \geq |a|$. For large x, the curves asymptotically approach straight lines, $y = \pm(b/a)x$. Of particular importance is the "unit" hyperbola, $x^2 - y^2 = 1$. The pairs (x, y) on this curve correspond to coshu and sinhu, respectively, the hyperbolic cosines and sines.

The formulas for the conics presented so far are in standard form, oriented so as to be symmetric about at least one axis. Naturally, it is possible to draw conic sections at skewed angles so that they do not exhibit symmetry about the axes. The penalty is that the x's and y's are not kept separate but form products. A simple example is $xy = 1$, which is a hyperbola with the axes as asymptotes. The center is at the origin with the major axis of length $2\sqrt{2}$, and the foci at $(-\sqrt{2}, -\sqrt{2})$ and $(\sqrt{2}, \sqrt{2})$.

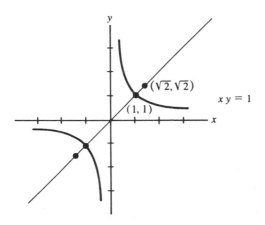

9.3 POLAR COORDINATES.

The rectangular coordinate system is very useful, but there are other geometries that make some problems more tractable. For a system in which rotational motion occurs, such as the solar system, a polar coordinate system might be more appropriate. This consists of an origin, a radius arm specifying the distance from the origin, and an angle specifying angular direction from an initial ray:

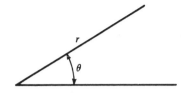

The description of a straight line in polar coordinates starts out with the specification of the radius line that is perpendicular to the line. The radius

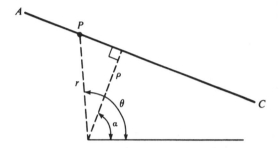

line of length ρ at angle α is perpendicular to the desired line AC. All other points on the line are given by

$$r \cos(\theta - \alpha) = \rho \qquad (9\text{--}40)$$

If the line goes through the origin, the equation is $\theta = K$. If the line is parallel to the y axis, $\alpha = 0$ and $r \cos \theta = \rho$. If the line is parallel to the x axis, $\alpha = 90°$ and $r \cos(\theta - 90°) = r \sin \theta = \rho$.

The general equation for a circle is more complicated in polar coordinates than in Cartesian. The center of the circle with radius a is at the fixed point

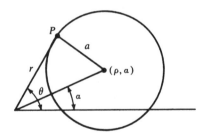

(ρ, α). Any point on the circle is given by (r, θ). From the law of cosines

$$r^2 + \rho^2 - 2r\rho \cos(\theta - \alpha) = a^2 \qquad (9\text{--}41)$$

Some special cases reduce this equation to a much simpler form.

If the circle is centered at the origin, (ρ, α) is $(0, 0)$ and

$$r = a \qquad (9\text{--}42)$$

If the circle is centered on the x axis with the polar origin on the circumference, (ρ, α) is $(a, 0)$ and

137

$$r = 2a \cos \theta \qquad (9\text{–}43)$$

If the circle is centered on the y axis with the polar origin on the circumference, (ρ, α) is $(a, 90°)$ and

$$r = 2a \sin \theta \qquad (9\text{–}44)$$

Conic sections can be completely described with just one definition in polar coordinates. The new definition says that a conic section is the locus of points whose distance from a point divided by its distance from a line is a constant. The line is called the directrix, the point is called the focus, and the ratio of distances is called the eccentricity, e.

Armed with this definition we can derive an expression for r in terms of θ. The distance from the focus equals r. The distance to the directrix from the point (r, θ) is $p + r \cos \theta$.

$$e = \frac{r}{p + r \cos \theta} \qquad (9\text{–}45)$$

$$r = ep + er \cos \theta$$

$$r = \frac{ep}{1 - e \cos \theta} \qquad (9\text{–}46)$$

This expression is frequently written in terms of r_0, the distance from the focus to the curve when the radius arm is parallel to the directrix.

$$e = \frac{r_0}{p + r_0 \cos (90°)} = \frac{r_0}{p} \qquad (9\text{–}47)$$

$$r_0 = ep$$

$$r = \frac{r_0}{1 - e \cos \theta} \qquad (9\text{–}48)$$

We said that e is a measure of the out-of-roundness of the figure, and indeed it is. If $e = 0$, then $r = r_0$, and we have a circle; the directrix must be at infinity.

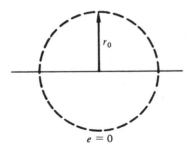

$e = 0$

For $0 \leq e \leq 1$, the figure becomes an ellipse. Note that e of an ellipse equals c/a just as before with Cartesian coordinates.

$$2a = \frac{r_0}{1 - e \cos 0} + \frac{r_0}{1 - e \cos \pi} = \frac{r_0}{1 - e} + \frac{r_0}{1 + e} = \frac{2r_0}{1 - e^2}$$

$$c = a - \frac{r_0}{1 + e} = \frac{r_0}{1 - e^2} - \frac{r_0 - er_0}{1 - e^2} = \frac{er_0}{1 - e^2}$$

Therefore,

$$\frac{c}{a} = e \qquad\qquad (9\text{--}49)$$

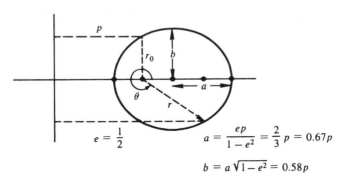

$e = \frac{1}{2}$

$$a = \frac{ep}{1 - e^2} = \frac{2}{3} p = 0.67p$$

$$b = a\sqrt{1 - e^2} = 0.58p$$

A parabola has eccentricity equal to one, the various shapes depending only on the distance from focus to directrix.

$r = 2p$

$\theta = 60°$

$e = 1$

$$p = r_0$$

$$r = \frac{p}{1 - \cos \theta} \qquad\qquad (9\text{--}50)$$

Parabolas with three different values of p are shown in the diagram.

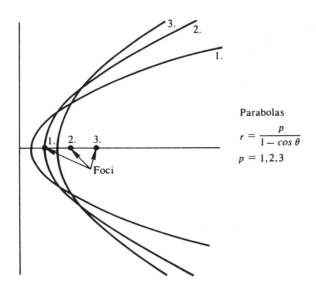

Parabolas

$$r = \frac{p}{1 - \cos \theta}$$

$$p = 1, 2, 3$$

With $e > 1$, one side of a hyperbola appears. A mirror on the directrix would produce the other half.

$$r_0 = 2p$$

$$r = \frac{r_0}{1 - 2 \cos \theta} \qquad (9\text{–}51)$$

Hyperbolas with three different values of eccentricity are shown in the diagram.

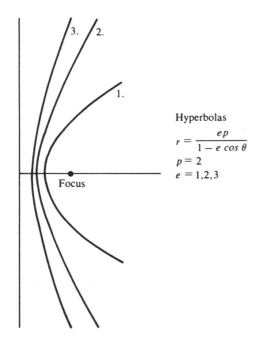

Hyperbolas

$$r = \frac{ep}{1 - e \cos \theta}$$

$p = 2$

$e = 1,2,3$

9.4 CYLINDRICAL COORDINATE SYSTEM.

The polar coordinate system extended to three dimensions is called the cylindrical coordinate system. Its use makes it easier to describe phenomena with cylindrical sources or boundaries.

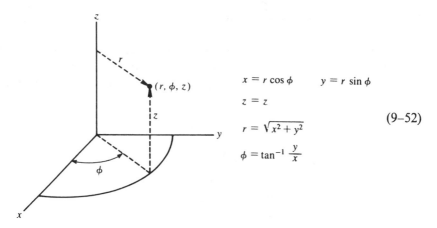

$$x = r \cos \phi \qquad y = r \sin \phi$$

$$z = z$$

$$r = \sqrt{x^2 + y^2}$$

$$\phi = \tan^{-1} \frac{y}{x}$$

$$(9\text{–}52)$$

A differential element of volume in Cartesian coordinates is simply

$$dV = dx\, dy\, dz \tag{9–53}$$

In cylindrical coordinates the situation is a little different. Note that dimensionally the differential elements must yield a product that is a length cubed.

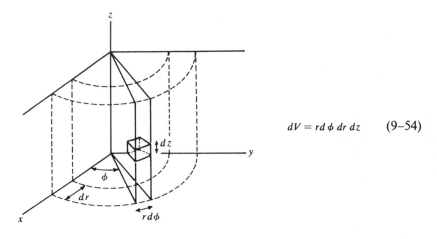

$$dV = rd\phi\, dr\, dz \qquad (9\text{-}54)$$

Vectors in cylindrical coordinates can be expressed in terms of magnitudes times unit vectors, $\hat{r}$, $\hat{\phi}$, and $\hat{k}$. The Cartesian unit vectors are $\hat{i}, \hat{j}$, and $\hat{k}$.

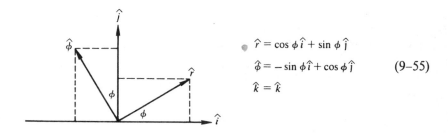

$$\hat{r} = \cos\phi\,\hat{i} + \sin\phi\,\hat{j}$$
$$\hat{\phi} = -\sin\phi\,\hat{i} + \cos\phi\,\hat{j} \qquad (9\text{-}55)$$
$$\hat{k} = \hat{k}$$

A change in the Cartesian components of a vector is in magnitude only; the unit vectors do not change direction. In cylindrical coordinates, a change in position makes the r vector change from one φ to another. It appears that

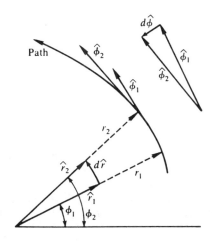

$d\hat{r}$ is in the $\hat{\varphi}$ direction, and $d\hat{\varphi}$ is in the negative $\hat{r}$ direction. This same result can be demonstrated analytically by differentiating directly the unit vectors $\hat{r}$ and $\hat{\varphi}$ in terms of their identities in $\hat{i}$ and $\hat{j}$. (Differentiation with respect to time is denoted by a dot over the symbol of the variable.)

$$\hat{r} = \cos\varphi\hat{i} + \sin\varphi\hat{j} \qquad \dot{\hat{r}} = -\dot{\varphi}\sin\varphi\hat{i} + \dot{\varphi}\cos\varphi\hat{j} = \dot{\varphi}\hat{\varphi}$$

$$\hat{\varphi} = -\sin\varphi\hat{i} + \cos\varphi\hat{j} \qquad \dot{\hat{\varphi}} = -\dot{\varphi}\cos\varphi\hat{i} - \dot{\varphi}\sin\varphi\hat{j} = -\dot{\varphi}\hat{r} \qquad (9\text{-}56)$$

Notice that in performing the differentiation, $\dot{\hat{i}} = 0$ and $\dot{\hat{j}} = 0$. The unit Cartesian vectors do not change direction, and since they are always of unit length, do not change length. In differentiating a vector that is expressed in terms of $\hat{r}$ and $\hat{\varphi}$, the values for $\dot{\hat{r}}$ and $\dot{\hat{\varphi}}$ must be taken into account.

$$\mathbf{r} = \rho\hat{r} + z\hat{k}$$

$$\mathbf{v} = \dot{\rho}\hat{r} + \rho\dot{\hat{r}} + \dot{z}\hat{k} = \dot{\rho}\hat{r} + \rho\dot{\varphi}\hat{\varphi} + \dot{z}\hat{k} \qquad (9\text{-}57)$$

$$\mathbf{a} = (\ddot{\rho} - \rho\dot{\varphi}^2)\hat{r} + (\rho\ddot{\varphi} + 2\dot{\rho}\dot{\varphi})\hat{\varphi} + \ddot{z}\hat{k}$$

The acceleration contains a centripetal term in the $\hat{r}$ direction ($-\rho\dot{\varphi}^2 = -\rho\omega^2$) and a Coriolis term in the $\hat{\varphi}$ direction ($2\dot{\rho}\dot{\varphi} = 2\dot{\rho}\omega$). This latter term produces acceleration perpendicular to the radius vector if there is velocity along the radius.

9.5 SPHERICAL COORDINATE SYSTEM. Spherical coordinates are usually easiest to use for problems involving spherical sources or boundaries.

$$x = r\sin\theta\cos\phi$$
$$y = r\sin\theta\sin\phi$$
$$z = r\cos\theta$$
$$r = \sqrt{x^2 + y^2 + z^2} \qquad (9\text{-}58)$$
$$\phi = \tan^{-1}\frac{y}{x}$$
$$\theta = \tan^{-1}\frac{\sqrt{x^2 + y^2}}{z}$$

The differential volume and surface area elements can be deduced from the diagram:

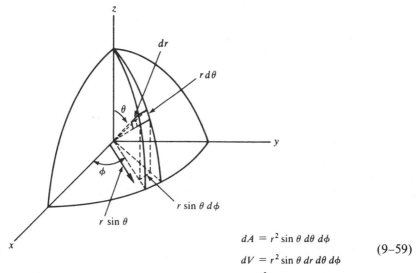

$$dA = r^2 \sin \theta \, d\theta \, d\phi$$
$$dV = r^2 \sin \theta \, dr \, d\theta \, d\phi$$
$$(9\text{–}59)$$

The unit vectors in spherical coordinates are $\hat{r}$, $\hat{\theta}$, and $\hat{\phi}$. Their relationship to the Cartesian unit vectors can be deduced from the diagrams:

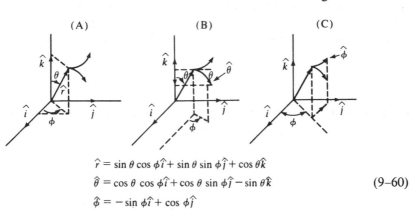

(A)   (B)   (C)

$$\hat{r} = \sin \theta \cos \phi \hat{i} + \sin \theta \sin \phi \hat{j} + \cos \theta \hat{k}$$
$$\hat{\theta} = \cos \theta \cos \phi \hat{i} + \cos \theta \sin \phi \hat{j} - \sin \theta \hat{k} \qquad (9\text{–}60)$$
$$\hat{\phi} = -\sin \phi \hat{i} + \cos \phi \hat{j}$$

Differentiating the spherical unit vectors produces a complex relationship among the three directions, but it is a relationship that yields useful insights into some physical phenomena.

$$\hat{r} = \sin \theta \cos \varphi \hat{i} + \sin \theta \sin \varphi \hat{j} + \cos \theta \hat{k}$$
$$\hat{\theta} = \cos \theta \cos \varphi \hat{i} + \cos \theta \sin \varphi \hat{j} - \sin \theta \hat{k}$$
$$\hat{\varphi} = -\sin \varphi \hat{i} + \cos \varphi \hat{j}$$
$$\dot{\hat{r}} = \dot{\theta} \cos \theta \cos \varphi \hat{i} - \dot{\varphi} \sin \theta \sin \varphi \hat{i} + \dot{\theta} \cos \theta \sin \varphi \hat{j} + \dot{\varphi} \sin \theta \cos \varphi \hat{j}$$
$$- \dot{\theta} \sin \theta \hat{k} = \dot{\theta}[\cos \theta \cos \varphi \hat{i} + \cos \theta \sin \varphi \hat{j} - \sin \theta \hat{k}]$$
$$+ \dot{\varphi}[\sin \theta \cos \varphi \hat{j} - \sin \theta \sin \varphi \hat{i}] = \underline{\dot{\theta} \hat{\theta} + \dot{\varphi} \sin \theta \hat{\varphi}}$$
$$(9\text{-}61)$$

$$\dot{\hat{\theta}} = -\dot{\theta}\sin\theta\cos\varphi\hat{i} - \dot{\varphi}\cos\theta\sin\varphi\hat{i} - \dot{\theta}\sin\theta\sin\varphi\hat{j} + \dot{\varphi}\cos\theta\cos\varphi\hat{j}$$
$$- \dot{\theta}\cos\theta\hat{k} = -\dot{\theta}[\sin\theta\cos\varphi\hat{i} + \sin\theta\sin\varphi\hat{j} + \cos\theta\hat{k}]$$
$$+ \dot{\varphi}\cos\theta[\cos\varphi\hat{j} - \sin\varphi\hat{i}] = \underline{-\dot{\theta}\hat{r} + \dot{\varphi}\cos\theta\hat{\varphi}}$$
$$\dot{\hat{\varphi}} = -\dot{\varphi}\cos\varphi\hat{i} - \dot{\varphi}\sin\varphi\hat{j} = \underline{-\dot{\varphi}\sin\theta\hat{r} - \dot{\varphi}\cos\theta\hat{\theta}}$$

$$\left(\begin{array}{l}\text{(This last identity is most easily proved by substituting for }\hat{r}\text{ and }\hat{\theta}.) \\ \sin\theta\hat{r} + \cos\theta\hat{\theta} = \sin^2\theta\cos\varphi\hat{i} + \sin^2\theta\sin\varphi\hat{j} + \sin\theta\cos\theta\hat{k} \\ \qquad\qquad + \cos^2\theta\cos\varphi\hat{i} + \cos^2\theta\sin\varphi\hat{j} - \sin\theta\cos\theta\hat{k} \\ \qquad\quad = \cos\varphi\hat{i} + \sin\varphi\hat{j}\end{array}\right)$$

Note that the time derivative of each of the unit vectors consists of components in the other two directions. Let us use these values to find the first and second time derivatives of the position vector in spherical coordinates.

$$\mathbf{r} = r\hat{r}$$
$$\mathbf{v} = \dot{r}\hat{r} + r\dot{\hat{r}} = \dot{r}\hat{r} + r\dot{\theta}\hat{\theta} + r\dot{\varphi}\sin\theta\hat{\varphi}$$
$$\mathbf{a} = [\ddot{r} - r\dot{\theta}^2 - r\dot{\varphi}^2\sin^2\theta]\hat{r} + [r\ddot{\theta} + 2\dot{r}\dot{\theta} - r\dot{\varphi}^2\sin\theta\cos\theta]\hat{\theta} \qquad (9\text{-}62)$$
$$+ [r\ddot{\varphi}\sin\theta + 2\dot{r}\dot{\varphi}\sin\theta + 2r\dot{\theta}\dot{\varphi}\cos\theta]\hat{\varphi}$$

This formula for acceleration contains the full panoply of effects for motion on and around the earth. The earth is, after all, a spherical object, and anyone riding on it already has a value for $\dot{\varphi}$. It is ω, the angular velocity of the earth, which has varying effects on the person depending on his geographical latitude ($90° - \theta$, since latitude is measured from the equator and θ is measured from the pole). Consider a simple, special case of motion on the earth's surface ($\dot{r} = 0$), moving along a meridian line ($\dot{\varphi} = \omega$, the earth's angular velocity, and $\ddot{\varphi} = 0$), with constant velocity ($\ddot{\theta} = 0$).

$$\mathbf{a} = [-r\dot{\theta}^2 - r(\omega\sin\theta)^2]\hat{r} + [-r\omega^2\sin\theta\cos\theta]\hat{\theta}$$
$$+ [2r\dot{\theta}\omega\cos\theta]\hat{\varphi} \qquad (9\text{-}63)$$

Consider first the effects on the equator where $\theta = 90°$. There is no component of acceleration in the $\hat{\theta}$ or $\hat{\varphi}$ direction. There are, however, two causes for centripetal acceleration in the $-\hat{r}$ direction. One is the centripetal acceleration caused by the rotation of the earth, $r\omega^2$, and the other is caused by the north–south velocity, which is also a rotational motion.

$$\mathbf{a} = -[r\dot{\theta}^2 + r\omega^2]\hat{r} \qquad (9\text{-}64)$$

At other latitudes, the centripetal acceleration produced by the earth's rotation is less; the term is $r(\omega\sin\theta)^2$ for the $\hat{r}$ component. There is also a $\hat{\theta}$ component equal to $-r\omega^2\sin\theta\cos\theta$, having its maximum amplitude at $\theta = 45°$. Besides these accelerations, which are independent of the velocity, $r\dot{\theta}$, a Coriolis effect appears, having its maximum amplitude at the poles: $[2r\dot{\theta}\omega\cos\theta]\hat{\varphi}$. Note that it is proportional to the velocity, $r\dot{\theta}$. For $\dot{\theta}$ positive, it is in the positive $\hat{\varphi}$ direction north of the equator (θ from 0 to 90°; cos θ

from $+1$ to 0), and in the negative $\hat{\varphi}$ direction south of the equator (θ from 90 to 180°; $\cos \theta$ from 0 to -1). These accelerations are the ones observed from outside the system; thus it is a *centripetal* force in the negative r direction, toward the center. The earthbound observer witnesses a *centrifugal* force, away from the center. Similarly, the Coriolis accelerations in our formulas are those seen by an outside observer. He would say that those accelerations must be present if the object continues to have the observed motion. Bound to the earth we would find that an object moving from north to south (positive $\dot{\theta}$) in the northern hemisphere would experience an acceleration to the right ($-\varphi$). Indeed, for any direction of motion along the earth's surface in the northern hemisphere an object will experience a force to the right. In the southern hemisphere, the force will be to the left. This Coriolis force is responsible for the counterclockwise rotation of storms with low pressure centers in the northern hemisphere.

The various analytic geometries presented here all have some elements in common. One is the unique representation of every point in the field it covers. Another is the use of mutually perpendicular directions to specify the location of each point. The choice of system is arbitrary, since they are all equivalent, but the one which is easiest to use will be the one that best fits the geometry of the sources and boundaries.

9.6 ORBITS AND SPECIAL CURVES

1. *Gravitational orbits.* The orbit of any object subject only to a central force (F_r) must be one of the conic figures. Which one depends on the relationship of its potential energy to its total energy.

As the first step in deriving the orbit, note that a central force imposes two special conditions. To find these, form the vector product of r and the two sides of Newton's second law:

$$\mathbf{r} \times \mathbf{F}_r = \mathbf{r} \times m\frac{d^2\mathbf{r}}{dt^2} \qquad (9\text{--}65)$$

The first side is zero since $\mathbf{r}$ and $\mathbf{F}_r$ are parallel if $\mathbf{F}_r$ is central (radial), and the vector product of parallel vectors is zero. We are left with

$$\mathbf{r} \times \frac{d^2\mathbf{r}}{dt^2} = 0 \qquad (9\text{--}66)$$

When integrated, this gives

$$\mathbf{r} \times \frac{d\mathbf{r}}{dt} = 2K \qquad (9\text{--}67)$$

To prove this, perform a differentiation of the integral

$$\frac{d}{dt}\left[\mathbf{r} \times \frac{d\mathbf{r}}{dt}\right] = \frac{d\mathbf{r}}{dt} \times \frac{d\mathbf{r}}{dt} + \mathbf{r} \times \frac{d^2\mathbf{r}}{dt^2} = 0$$

and $(d\mathbf{r}/dt) \times (d\mathbf{r}/dt) = 0$, necessarily.

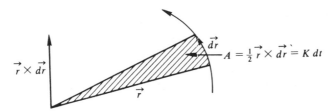

$$\vec{A} = \tfrac{1}{2}\vec{r} \times \vec{dr} = K\,dt$$

The integral is equal to twice the area swept out by the radius arm. This confirms Kepler's second law that the radius arm to any of the planets sweeps out equal areas in equal times $\mathbf{r} \times d\mathbf{r} = 2K\,dt$. Furthermore, since $\mathbf{r} \times d\mathbf{r}$ defines a normal to the area, and the normal stays constant, the orbit must lie in a plane. [This is also a proof that angular momentum is conserved, since $m\mathbf{r} \times (d\mathbf{r}/dt)$ is angular momentum.]

The planet must conserve energy. (We assume that the mass of the controlling body is so large that only the planetary motion and energy need be considered.) For an inverse-square force, the potential energy is inversely proportional to the radius. The conservation of energy equation is therefore

$$\frac{1}{2}mv^2 - G\frac{mM}{r} = \frac{1}{2}mv_0^2 - G\frac{mM}{r_0} \tag{9-68}$$

Transform this equation and the law of areas to polar coordinates and eliminate t, using $v^2 = \dot{r}^2 + r^2\dot{\varphi}^2$; $d/dt = \dot{\varphi}(d/d\varphi)$.

(1) $\quad \dfrac{1}{2}m(\dot{r}^2 + r^2\dot{\varphi}^2) - G\dfrac{mM}{r} = \dfrac{1}{2}mv_0^2 - G\dfrac{mM}{r_0}$

(2) $\quad \mathbf{r} \times \dfrac{d\mathbf{r}}{dt} = 2K$

(1') $\quad \dfrac{1}{2}m\dot{\varphi}^2\left(\dfrac{dr}{d\varphi}\right)^2 + \dfrac{1}{2}mr^2\dot{\varphi}^2 - G\dfrac{mM}{r} = \dfrac{1}{2}mv_0^2 - G\dfrac{mM}{r_0}$

(2') $\quad \mathbf{r} \times \dfrac{d\mathbf{r}}{dt} = \mathbf{r} \times \left|\dfrac{dr}{dt}\right|\hat{r} + \mathbf{r} \times |r|\dfrac{d\hat{r}}{dt} = |r^2\dot{\varphi}| = 2K \qquad (\dot{\hat{r}} = \dot{\varphi}\hat{\varphi})$

[This product can also be deduced in terms of the triangular area in the diagram: $(\mathbf{r} \times r\dot{\boldsymbol{\varphi}})$.]

(1' and 2') $\quad \dfrac{4K^2}{r^4}\left(\dfrac{dr}{d\varphi}\right)^2 + \dfrac{4K^2}{r^2} - \dfrac{2GM}{r} = v_0^2 - \dfrac{2GM}{r_0}$

The variables r and φ can be separated to yield

$$d\varphi = \frac{(2K/r^2)\,dr}{\sqrt{[v_0^2 - (2GM/r_0)] + (2GM/r) - (4K^2/r^2)}}$$

The integral is easier to interpret if we substitute $u = 1/r$ and $du = -(1/r^2)\,dr$:

$$\varphi + c = \int \frac{-2K\,du}{\sqrt{[v_0^2 - (2GM/r_0)] + 2GMu - 4K^2u^2}}$$

The integral is an arc cosine form:

$$-\int \frac{dx}{\sqrt{a + 2bx - cx^2}} = \frac{1}{\sqrt{c}} \cos^{-1}\left(\frac{b - cx}{\sqrt{b^2 + ac}}\right)$$

$$\cos(\varphi + c) = \frac{GM - 4K^2u}{\sqrt{G^2M^2 + [v_0^2 - (2GM/r_0)]4K^2}}$$

$$r = \frac{4K^2}{GM - \sqrt{G^2M^2 + [v_0^2 - (2GM/r_0)]4K^2}\ \cos(\phi + c)}$$

$$r = \frac{r_0}{1 - \sqrt{1 + [v_0^2/G^2M^2) - (2/r_0GM)]4K^2}\ \cos\phi}$$

(9-69)

where $r_0 = 4K^2/GM$ and c is chosen to equal 0.

This is the standard form for a conic curve in polar coordinates. The eccentricity is

$$e = \sqrt{1 + \left(\frac{v_0^2}{G^2M^2} - \frac{2}{r_0GM}\right)4K^2} = \sqrt{1 + \left(\frac{1}{2}mv_0^2 - \frac{GmM}{r_0}\right)\frac{8K^2}{G^2M^2m}}$$

(9-70)

If $e < 1$, the orbit is an ellipse; if $e > 1$, the orbit is hyperbolic. In particular, if $\frac{1}{2}mv_0^2 = G(mM/r_0)$, the total energy is zero (positive kinetic energy equals negative potential energy). The eccentricity is 1, and the orbit is parabolic. If the kinetic energy is less than the potential $[\frac{1}{2}mv_0^2 < G(mM/r_0)]$, $e < 1$ and the planet is bound in an elliptical orbit. If kinetic energy is greater than the potential, the planet is on a hyperbolic path and is not bound.

There is a special case when the kinetic energy is equal to one half the potential energy $[\frac{1}{2}mv_0^2 = \frac{1}{2}G(mM/r_0)]$.

$$\sqrt{1 + \left(\frac{1}{2}mv_0^2 - G\frac{mM}{r_0}\right)\frac{8K^2}{G^2M^2m}}$$

$$= \sqrt{1 - \frac{1}{2}\frac{GMm}{r_0}\frac{8K^2}{G^2M^2m}} = \sqrt{1 - \frac{4K^2}{r_0GM}} = \sqrt{1 - \frac{4r_0GM/4}{r_0GM}}$$

$$= \sqrt{1 - 1} = 0$$

The eccentricity is zero; the orbit is a circle, which has the smallest ratio possible of kinetic to potential energy.

2. The Catenary. A rope hangs under its own weight. The downward force for each segment of length is proportional to the arc length, $ds = \sqrt{dx^2 + dy^2}$. The horizontal component of tension throughout the rope must be constant, since no segment of the rope is moving sideways.

GEOMETRY

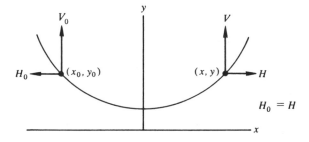

The length of the rope between the two points is

$$L = \int_{x_0}^{x} \sqrt{1 + \left(\frac{dy}{dx}\right)^2}\, dx$$

If the weight of the rope per unit length is ρ,

$$V + V_0 = \rho \int_{x_0}^{x} \sqrt{1 + \left(\frac{dy}{dx}\right)^2}\, dx$$

Divide by the constant H, noting that $V/H = dy/dx$:

$$\frac{dy}{dx} + \frac{V_0}{H} = \frac{\rho}{H} \int_{x_0}^{x} \sqrt{1 + \left(\frac{dy}{dx}\right)^2}\, dx$$

Differentiate both sides, setting $dy/dx = y'$:

$$y'' = \frac{\rho}{H}\sqrt{1 + y'^2}$$

Let $y' = r$:

$$r' = \frac{\rho}{H}\sqrt{1 + r^2}$$

The integral of this is

$$r = \sinh\left(\frac{\rho}{H}x + C_1\right) = \frac{dy}{dx}$$

The integral of this is:

$$y = \frac{H}{\rho}\cosh\left(\frac{\rho}{H}x + C_1\right) + C_2$$

If $y = 0$ and $dy/dx = 0$, when $x = 0$, $C_1 = 0$ and $C_2 = -(H/\rho)$:

$$y = \frac{H}{\rho}\left(\cosh\frac{\rho}{H}x - 1\right) = \frac{H}{2\rho}[e^{(\rho/H)x} + e^{-(\rho/H)x} - 2] \qquad (9\text{--}71)$$

3. The Cycloid. The cycloid is generated by tracing a point on the circumference of a circle as the circle rolls along a straight line.

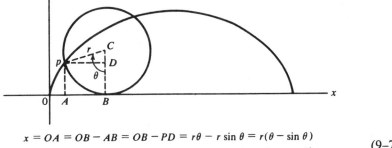

$$x = OA = OB - AB = OB - PD = r\theta - r\sin\theta = r(\theta - \sin\theta)$$
$$y = AP = BD = BC - CD \qquad = r\ -r\cos\theta = r(1-\cos\theta)$$

(9–72)

The cycloid has a couple of interesting dimensions associated with it. The length of the arch is equal to eight times the radius of the rolling circle, and the area enclosed between the arch and the straight line is equal to $3\pi r^2$.

When turned upside down, the cycloid is the path of quickest descent of falling from any point to any point. Indeed, the descent time is not only less for the cycloid than for any other path, the descent time on the cycloid is the same from any point to the same point!

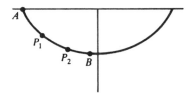

Time A–B for friction-free fall is faster than for any other path and also equals the time for P_1–B or P_2–B. Another name for the curve having this property is *brachistochrone*.

9.7 PYTHAGOREAN THEOREM. One property of right triangles is so frequently used that we record here its geometrical proof.

$$c^2 = a^2 + b^2 \qquad\qquad (9\text{–}73)$$

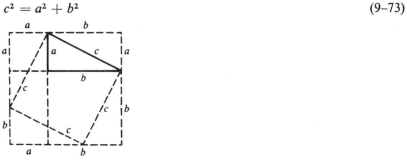

GEOMETRY

Consider the areas of the squares, rectangles, and triangles formed by the projections of sides a, b, and c.

$$a^2 + b^2 + ab + ab = c^2 + \tfrac{1}{2}ab + \tfrac{1}{2}ab + \tfrac{1}{2}ab + \tfrac{1}{2}ab$$

area of large square = area of central square + 4 triangular regions
Therefore,

$$a^2 + b^2 = c^2$$

9.8 AREAS AND VOLUMES OF VARIOUS GEOMETRICAL FIGURES

Circle: circumference $= 2\pi r = \pi d$ area $= \pi r^2 = \pi \dfrac{d^2}{4}$

Rectangle: perimeter $= 2(a+b)$ area $= ab$

Parallelogram: perimeter $= 2(a+b)$ area $= hb = (a \sin \theta)b$

Triangle: perimeter $= a+b+c$ area $= \dfrac{1}{2}hb = \dfrac{1}{2}(a \sin \theta)b$

Ellipse: perimeter $= 2\pi a\left(1 - \dfrac{1}{4}e^2 - \dfrac{3}{64}e^4 \ldots\right) \approx \pi\sqrt{2(a^2+b^2)}$ area $= \pi ab$

Areas of regular polygons with n sides, each of length s: $A = s^2 \dfrac{n\cot \dfrac{180°}{n}}{4}$

Name	n	$\dfrac{1}{4} n\cot \dfrac{180°}{n}$
Triangle	3	0.433
Square	4	1.000
Pentagon	5	1.720
Hexagon	6	2.598
Heptagon	7	3.634
Octagon	8	4.828
Decagon	10	7.694
Dodecagon	12	11.196

Sphere: surface area $= 4\pi r^2 = \pi d^2$ volume $= \dfrac{4}{3}\pi r^3 = \dfrac{1}{6}\pi d^3$

Cylinder: curved surface area $= 2\pi rh$ volume $= \pi r^2 h$

Cone: curved surface area $= \pi rl$ for right circular cone; l is distance from vertex to point on circumference at base   volume $= \dfrac{1}{3}Ah$ where A is area of base, and h is vertical height from base to vertex

Ellipsoid: area is approximately $= \dfrac{1}{3}\pi b\left[7a + b\left(6-\dfrac{b}{a}\right)\right]$ volume $= \dfrac{4}{3}\pi ab^2$

Prism: volume of prism with parallel sides and any shape of base $= hA_{\text{base}}$, where h is the vertical height from base to base

The regular solids:

Name	Number of faces	Area = edge² ×	Volume = edge³ ×
Tetrahedron	4	1.732	0.118
Cube	6	6.000	1.000
Octahedron	8	3.464	0.471
Dodecahedron	12	20.646	7.663
Icosahedron	20	8.660	2.182

CHAPTER10 VECTORS

Vectors represent physical quantities that have both magnitude and direction, but not all such quantities can be represented by vectors. It is also necessary for these quantities to combine according to vector rules. Ordinary forces, for instance, have magnitude and direction, and do combine to produce net forces in the same way that the vector model predicts. Stresses, however, are usually more complicated. They have magnitude and direction, but must be described in terms of tensors. Vector properties, including an axial direction, can be assigned to angular velocity, and the rules of combination are satisfied. The same description does not work, however, for angular displacement. Mountains have magnitude and direction (up), but cannot be represented by vectors. On paper, vectors are usually pictured by drawing arrows, but as any Indian will testify, arrows cannot be decomposed into components like vectors. Whether or not a physical quantity has vector properties must usually be determined by experiment.

10.1 ADDITION AND SUBTRACTION OF VECTORS. The basic vector properties are defined in terms of the properties of displacements in space. A displacement is not just a distance, but also is characterized by a direction: 3 miles south; 2 m straight up; 4 cm along the x axis. Such displacements can be represented by arrows drawn on paper. A displacement (as

opposed to an actual journey) is independent of the starting point. It is characterized completely by a magnitude and a direction. The arrow representing it can be shifted about on the paper to combine with other displacements, so long as its magnitude and direction remain unchanged. Displacements can be added by determining the net distance and direction from the starting point after a series of displacements. The addition is commutative; i.e., $A + B = B + A$.

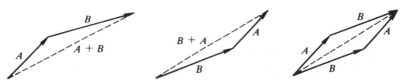

Vector addition can be represented by combining the arrows, head to tail. The negative of a displacement has the same magnitude as the positive displacement but is in the opposite direction.

The subtraction of B from A is consequently just the addition of A and $-B$.

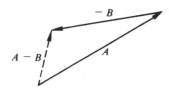

As a practical matter, there is hardly ever an excuse for performing these operations by drawing scale arrows on paper. Instead, the displacements are broken down into perpendicular components. The algebraic sum of the components along one axis is equal to the component along that axis of the displacement sum.

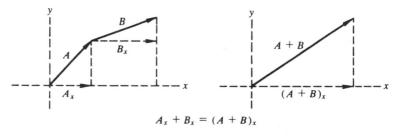

$$A_x + B_x = (A + B)_x$$

For example, a man walks 1 mile east, 30° N of E for 2 miles, 2 miles W, and 2 miles SW (45° S of W). Where does he end up?

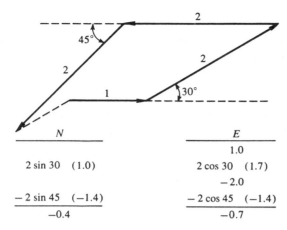

N	E
	1.0
2 sin 30 (1.0)	2 cos 30 (1.7)
	−2.0
− 2 sin 45 (−1.4)	− 2 cos 45 (−1.4)
−0.4	−0.7

The resultant displacement has a north component of −0.4 (−N is +S), and an east component of −0.7 (−E is +W). The magnitude of this displacement is equal to $\sqrt{(0.4)^2 + (0.7)^2} = 0.8$ mile. The direction is such that $\tan \theta = 0.7/0.4 = 1.75$. The direction from the starting point is approximately 60° W of S.

Any quantity that can be combined or decomposed in the same way as displacements can be represented by vectors. The common examples are velocity (as opposed to speed), acceleration, force, angular velocity, and torque. There are some problems even with these. Velocity, for instance, does not follow the combination rules when speeds are high enough to require the use of the special theory of relativity. Torque and angular momentum are technically *pseudo-vectors* because they do not reverse sign when the polarity of all their coordinates is changed. [$A(x, y, z) = -A(-x, -y, -z)$ for displacements, but $T(x, y, z) = +T(-x, -y, -z)$ for torques.]

Quantities that can be described in terms of magnitude only (though sometimes with the addition of + or −, indicating a value greater or less than some zero value) are called *scalars*. Common examples are speed (simply how fast, without regard to direction), voltage, temperature, length (as opposed to displacement), mass (so far as measurements now indicate, inertial mass is the same whether motion is directed toward the center of our galaxy or perpendicular to it), time, and volume (though not area, as we shall see).

A convenient way to describe vectors is in terms of multiples of unit vectors along perpendicular axes. The standard notation in Cartesian coordinates is shown in the diagram.

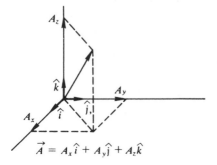

$$\vec{A} = A_x \hat{i} + A_y \hat{j} + A_z \hat{k} \qquad (10-1)$$

The components of vector A are given in terms of a scalar coefficient times a unit vector in each of the three perpendicular directions. The sum of the two vectors is

$$\mathbf{A} + \mathbf{B} = (A_x + B_x)\hat{i} + (A_y + B_y)\hat{j} + (A_z + B_z)\hat{k} \qquad (10\text{-}2)$$

The general rule for addition or subtraction of vectors is the same as we used in the example of finding the resultant of a series of displacements. Resolve each vector into its (x, y, z) components. Add the coefficients of each unit vector separately. The magnitude of the resultant is equal to

$$\sqrt{(A_x + B_x)^2 + (A_y + B_y)^2 + (A_z + B_z)^2} \qquad (10\text{-}3)$$

10.2 MULTIPLICATION OF VECTORS. There are several vector products that serve as useful descriptions of physical phenomena. The first involves multiplication of a vector by a scalar; e.g., force equals charge times field ($\mathbf{F}_{elec} = q\mathbf{E}$) or ($\mathbf{F}_{grav} = m\mathbf{g}$). A scalar–vector product merely changes the magnitude of the vector and not the direction. Thus the scalar multiplies each component of the vector. $\mathbf{F}_{elec} = q\mathbf{E} = qE_x\hat{i} + qE_y\hat{j} + qE_z\hat{k}$.

Another vector product involves the multiplication of two vectors. It is called the *dot* or *scalar* product because it is written $A \cdot B$ and results in a scalar.

$$\mathbf{A} \cdot \mathbf{B} = |A||B|\cos\theta \qquad (10\text{-}4)$$

where $|A|$ and $|B|$ are the magnitudes of $\mathbf{A}$ and $\mathbf{B}$, and θ is the angle between them. This product can be thought of as the projection of $\mathbf{B}$ on $\mathbf{A}$ times the magnitude of $\mathbf{A}$, or vice versa.

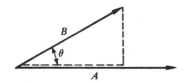

To find the scalar product you can also multiply like components together and add them. You can see that this is equivalent to the first formula by breaking the vectors into separate components and taking the dot products of all components. Since the unit vectors are all perpendicular, $\cos\theta$ will equal zero for all the products except for $\hat{i}\cdot\hat{i}$, $\hat{j}\cdot\hat{j}$, and $\hat{k}\cdot\hat{k}$. $\cos\theta$ will equal $+1$ or -1 for these depending on whether the like coefficients have the same sign or are of opposite sign.

$$
\begin{aligned}
\mathbf{A}\cdot\mathbf{B} &= (A_x\hat{i} + A_y\hat{j} + A_z\hat{k})\cdot(B_x\hat{i} + B_y\hat{j} + B_z\hat{k})\\
&= A_xB_x\hat{i}\cdot\hat{i} + A_xB_y\hat{i}\cdot\hat{j} + A_xB_z\hat{i}\cdot\hat{k} + A_yB_x\hat{j}\cdot\hat{i} + A_yB_y\hat{j}\cdot\hat{j}\\
&\quad + A_yB_z\hat{j}\cdot\hat{k} + A_zB_x\hat{k}\cdot\hat{i} + A_zB_y\hat{k}\cdot\hat{j} + A_zB_z\hat{k}\cdot\hat{k}\\
&= A_xB_x + A_yB_y + A_zB_z = |A||B|\cos\theta \qquad (10\text{-}5)
\end{aligned}
$$

VECTORS

Notice that the order of multiplication doesn't matter, since $\cos(-\theta) = \cos\theta$.

The work done on an object by a force **F** is the product of the force and the distance **s** it moves the object in the direction of **F**. This is a scalar product: $W = \mathbf{F}\cdot\mathbf{s}$. Similarly, the power, which is the work per unit time, is

$$p = \frac{\Delta W}{\Delta t} = \frac{\mathbf{F}\cdot\Delta\mathbf{s}}{\Delta t} = \mathbf{F}\cdot\mathbf{v}$$

An electric dipole consists of a negative charge separated by a small distance from a positive charge. When placed in an electric field, the dipole tends to line up with the field. The potential energy of the dipole in the field is $U = -\mathbf{E}\cdot\mathbf{p}$. The electric field is **E**, and **p** is the dipole moment, represented by a vector running from the negative charge to the positive and having a magnitude equal to the product of charge and separation distance between poles. The same formula applies to a magnet in a magnetic field.

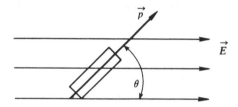

The other vector product that has widespread applications is the *cross product* or *vector product*, written $\mathbf{A} = \mathbf{B} \times \mathbf{C}$. In this case, the product, **A**, is a vector and has a direction perpendicular to both **B** and **C**. The magnitude of **A** is equal to $|B||C|\sin\theta$, where θ is the angle between **B** and **C** measured from **B**. The method of measuring the angle is important because

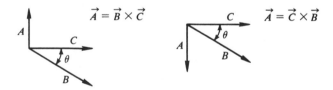

of the $\sin\theta$ term. If we measured the angle from **C**, we would get $-\theta$, and $\sin(-\theta) = -\sin\theta$. The order of multiplication, then, determines the product: $\mathbf{B} \times \mathbf{C} \neq \mathbf{C} \times \mathbf{B}$. Vector multiplication is not commutative. If the order is reversed, so is the sign of the product:

$$\mathbf{B} \times \mathbf{C} = -\mathbf{C} \times \mathbf{B} \qquad (10\text{–}6)$$

The direction of the perpendicular vector product is given by the right-hand rule. Think of the two vectors, *B* and *C*, lying in a plane. Rotate the first *clockwise*, in the same direction as you would to advance a right-hand screw; the direction of the product is the direction of the advancing screw.

10.2 MULTIPLICATION OF VECTORS

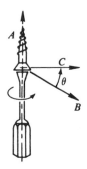

There is a method that uses determinants for finding the actual x, y, z components of the vector product (see p. 116). The rule is

$$\mathbf{A} = \mathbf{B} \times \mathbf{C} = \begin{vmatrix} \hat{i} & \hat{j} & \hat{k} \\ B_x & B_y & B_z \\ C_x & C_y & C_z \end{vmatrix} = \begin{matrix} \hat{i}(B_yC_z - B_zC_y) + \hat{j}(B_zC_x - B_xC_z) \\ + \hat{k}(B_xC_y - B_yC_x) \end{matrix} \quad (10\text{-}7)$$

Note that changing the order of multiplication to $C \times B$ would interchange the second and third row, thereby changing the sign of the product, a necessary consequence. The same result can be obtained by straightforward multiplication of the components of the two vectors. It is necessary to take into account the vector products of the unit vectors. The vector product of a vector and itself is zero, since the angle between them is zero. The vector product of any two unit vectors is $+$ or $-$ the third unit vector, since the product must be perpendicular to the first two vectors.

$$\hat{i} \times \hat{i} = \hat{j} \times \hat{j} = \hat{k} \times \hat{k} = 0$$
$$\hat{i} \times \hat{j} = -\hat{j} \times \hat{i} = \hat{k} \qquad \hat{j} \times \hat{k} = -\hat{k} \times \hat{j} = \hat{i}$$
$$\hat{k} \times \hat{i} = -\hat{i} \times \hat{k} = \hat{j} \qquad (10\text{-}8)$$

$$\begin{aligned} \mathbf{A} = \mathbf{B} \times \mathbf{C} &= (B_x\hat{i} + B_y\hat{j} + B_z\hat{k}) \times (C_x\hat{i} + C_y\hat{j} + C_z\hat{k}) \\ &= B_xC_x\hat{i} \times \hat{i} + B_xC_y\hat{i} \times \hat{j} + B_xC_z\hat{i} \times \hat{k} \\ &\quad + B_yC_x\hat{j} \times \hat{i} + B_yC_y\hat{j} \times \hat{j} + B_yC_z\hat{j} \times \hat{k} \\ &\quad + B_zC_x\hat{k} \times \hat{i} + B_zC_y\hat{k} \times \hat{j} + B_zC_z\hat{k} \times \hat{k} \\ &= \hat{i}(B_yC_z - B_zC_y) + \hat{j}(B_zC_x - B_xC_z) + \hat{k}(B_xC_y - B_yC_x) \quad (10\text{-}9) \end{aligned}$$

There are many physical phenomena that can be described using vector products. An electrically charged particle moving in a magnetic field experiences a force proportional to its velocity and the strength of the field, but perpendicular to both: $\mathbf{F} = q\mathbf{v} \times \mathbf{B}$. Angular momentum and torque are also vector cross products. They are, respectively, $\mathbf{L} = \mathbf{r} \times m\mathbf{v}$, and $\mathbf{T} = \mathbf{r} \times \mathbf{F}$. Since they are vector products of vectors, it is apparent why they are pseudo-vectors instead of true vectors. If the polarity of all coordinates is reversed (reflection through the origin), the primitive vectors $\mathbf{r}$, $\mathbf{v}$, and $\mathbf{F}$ will all change sign. The product of any two of them will therefore not change sign.

VECTORS

Consider how angular momenta combine in the case of a spinning top. Assume that as we look down at the top it is spinning counterclockwise. The direction of its angular momentum is therefore upward along the axis. Gravity exerts a torque whose vector direction is horizontal. The torque impulse creates a change of angular momentum, but the change is sideways along the horizontal rather than down. The top does not tip over; it precesses with angular velocity ω.

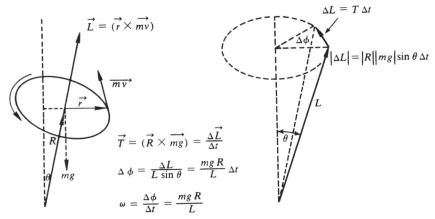

$$\vec{L} = (\vec{r} \times \overrightarrow{mv})$$

$$\Delta L = T \, \Delta t$$

$$|\Delta L| = |R||mg|\sin \theta \, \Delta t$$

$$\vec{T} = (\vec{R} \times \overrightarrow{mg}) = \frac{\overrightarrow{\Delta L}}{\Delta t}$$

$$\Delta \phi = \frac{\Delta L}{L \sin \theta} = \frac{mg\,R}{L} \, \Delta t$$

$$\omega = \frac{\Delta \phi}{\Delta t} = \frac{mg\,R}{L}$$

It is often useful to represent area as a vector. If we have two vectors **A** and **B**, they determine a parallelogram. Since the area of each triangle is

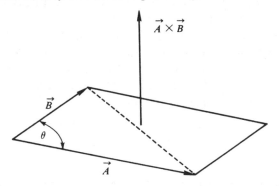

$$\vec{A} \times \vec{B}$$

$\frac{1}{2}|A||B|\sin \theta$, the magnitude of the area of the parallelogram is $|A||B|\sin \theta$. We can represent the area as the vector product $\mathbf{A} \times \mathbf{B}$, with a direction perpendicular to the surface.

The reasonableness of assigning vector properties to area can be seen by analyzing the flux of some influence (such as light or electric fields) through a given area. The "flow" of the influence depends not only on the strength of the influence and the magnitude of the area through which it is passing, but also on the angle between the flow and the area. In Faraday's conception of lines of force, the electric field is equal to the number of lines of force per unit area *perpendicular* to the lines. $\mathbf{E} = N/A$. In the standard

10.2 MULTIPLICATION OF VECTORS

symbols for electric fields and flux, $\Phi = \mathbf{E} \cdot \mathbf{\Delta S}$. The element of area, $\mathbf{\Delta S}$, is treated as a vector, and the flux, Φ, is the dot product of field and area.

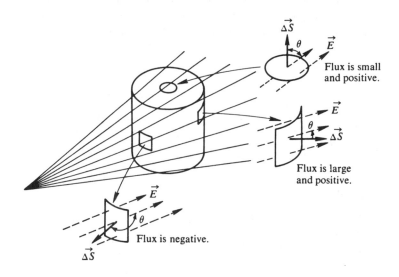

SAMPLES

1. A car drives 10 miles W, 30 miles NE ($45°$ N of E), 15 miles S, and 20 miles at $30°$ S of W. Where does it end up with respect to the starting point? (-3.8 N and -6.1 E, or 7.2 miles at an angle of $32°$ S of W)

2. A plane flies so that its ground speed is 470 miles/hr due east, while the wind is blowing at 180 miles/hr from the northeast ($45°$). What is the air speed of the plane and what is its heading?   (To counteract the head wind, it must have an air speed of 610 miles/hr, heading $12°$ N of E)

3. Find the dot products of the following vectors:

    (a) $\mathbf{A} = 4\hat{i} + 3\hat{j} + 6\hat{k}$ $\mathbf{B} = 2\hat{i} - \hat{j} - 4\hat{k}$ $(\mathbf{A} \cdot \mathbf{B} = -19)$

    (b) $\mathbf{F} = 2\hat{j} - \hat{k}$ $\mathbf{s} = 4\hat{i} + \hat{k}$ $(\mathbf{F} \cdot \mathbf{s} = -1)$

    (c) $\mathbf{E} = \hat{i} + \hat{j} + \hat{k}$ $\mathbf{p} = 3\hat{i} + 6\hat{j} - 9\hat{k}$ $(\mathbf{E} \cdot \mathbf{p} = 0)$

(Notice that the scalar product in the third problem is zero. This means that the two vectors are at right angles or *orthogonal*. The dot product of two orthogonal vectors is always zero.)

4. What is the torque on a door if a force of 40 N is applied at 60 cm from the hinges at an angle of $60°$ to the perpendicular?   (12 m $\cdot$ N.)

5. Find the cross products of the following vectors:

    (a) $\mathbf{A} = 4\hat{i} + 3\hat{j} + 6\hat{k}$

       $\mathbf{B} = 2\hat{i} - \hat{j} - 4\hat{k}$ $(\mathbf{A} \times \mathbf{B} = -6\hat{i} + 28\hat{j} - 10\hat{k}.)$

    (b) $\mathbf{F} = 2\hat{i} + 3\hat{j}$

       $\mathbf{r} = 3\hat{i} - 2\hat{j}$ $(\mathbf{r} \times \mathbf{F} = 13\hat{k})$

(Notice that **r** and **F** are both in the x, y plane. Therefore, their vector product is in the z direction.)

(c) $\mathbf{F} = 6\hat{i} + 4\hat{j}$

$\mathbf{r} = 3\hat{i} + 2\hat{j}$ $(\mathbf{r} \times \mathbf{F} = 0)$

(Notice here that **r** and **F** are in the x, y plane and parallel to each other. Therefore, their vector product is zero.)

CHAPTER11 COMPLEX NUMBERS

Complex numbers, in close analogy with the negatives, rationals, and irrational numbers, are an extension of the whole number system. They make it possible to perform all the standard algebraic operations on all numbers—such as finding the square root of a negative number. Despite their forbidding name, they are not "complex" and they should not be relegated to second-class numbership. Very real physical phenomena, which would be difficult or impossible to describe using only real numbers, can be rendered in concise and particularly simple form using complex numbers.

11.1 ADDITION AND MULTIPLICATION OF COMPLEX NUMBERS. Complex numbers consist of a "real" and an "imaginary" part. The two most common conventions for writing them are $a + bi$, where $i = \sqrt{-1}$ (hence the unfortunate term "imaginary") and a and b are real, and (a, b), where it is understood that a (multiplied by 1) is the real part and b (multiplied by i) forms the imaginary part. Both of these forms are instructive for envisioning a representation of complex numbers. Instead of using a number line, we must go to a plane with the y axis being the imaginary axis. $a + bi$ could be thought of as a vector in this plane, and (a, b) could be considered the coordinates:

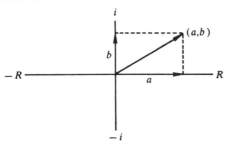

Notice that the real and imaginary parts are *orthogonal* and hence there is no mixing of the real and imaginary parts. (The concept of orthogonality comes up often in the study of vector spaces and linear algebra. In this case it is equivalent to the two axes being at right angles to each other.) This is important when adding two complex numbers. The reals are added together and the imaginary numbers are added together:

$$a + bi + c - di = a + c + (b - d)i$$
$$(a, b) + (c, -d) = (a + c, b - d) \tag{11-1}$$

When multiplying or dividing, however, every part must multiply both the real and imaginary parts of the other number. Remembering that $i^2 = -1$, we have

$$(a + bi) \times (c + di) = (ac - bd) + (ad + bc)i \tag{11-2}$$
$$(a, b) \times (c, d) = (ac - bd, ad + bc)$$

To divide $(a + bi)/(c + di)$ you should multiply both the top and bottom of the fraction by the complex conjugate of $c + di$, which is $(c + di)^* = (c - di)$. This eliminates all middle terms in the denominator, making it a real number. Doing this we have

$$\frac{a + bi}{c + di} = \frac{(a + bi)(c - di)}{c^2 + d^2} = \frac{(ac + bd) + (bc - ad)i}{c^2 + d^2} \tag{11-3}$$

Notice that $c^2 + d^2$ is just the sum of the squares of the legs of the right triangle in the complex plane and by the Pythagorean Theorem is just the square of the hypotenuse. It is the square of the magnitude of the complex number $c + di$ or, similarly, $c - di$.

11.2 THE COMPLEX EXPONENTIAL. Complex numbers can be raised to complex powers and the trigonometric, log, and exponential functions all have values for complex arguments. Most of these do not have simple physical significance. However, there is one that is extremely useful and we shall describe it here. It is $e^{i\theta}$. By power series or simple differentiation and substitution (see p. 194) you can show that

$$e^{i\theta} = \cos \theta + i \sin \theta \tag{11-4}$$

A complex number, then, could be described by giving the polar coordinates (r, θ), where

$$a + bi = re^{i\theta} = r \cos \theta + ir \sin \theta \tag{11-5}$$

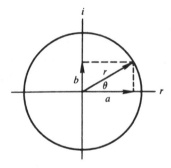

To show that a polar representation of the complex number system is useful, here is a method for solving equations such as $x^n = a + bi$. Let's do $x^3 = 1$, with $a = 1$, $b = 0$, and $n = 3$. This is a cubic, so it must have three roots. One obvious cube root of $+1$ is simply $+1$, but there must be two others. Putting $x = re^{i\theta}$, we have $x^3 = r^3 e^{i3\theta} = r^3 (\cos 3\theta + i \sin 3\theta) = 1$. The real parts on both sides of the equation must be equal, as must the imaginaries, which yields $r^3 \cos 3\theta = 1$ and $r^3 \sin 3\theta = 0$, or $\tan 3\theta = b/a = 0$. The tangent of 3θ equals zero when $\theta = 0$. Notice, however, that 0 is not the only angle that makes $\tan 3\theta = 0$. The others are $\pi/3$, $2\pi/3$, π, $4\pi/3$, and $5\pi/3$. There are others, of course, but they are all equivalent to one of these. For instance, $6\pi/3 = 2\pi$ is the same angle as 0; we have come full circle.

We can get r by either one of the equations $r^3 \cos 3\theta = 1$ or $r^3 \sin 3\theta = 0$. In this case the second one is of no help, but for $\theta = 0$ the first gives $r^3 = 1$, $r = 1$, the principal root. For the angles $\frac{2}{3}\pi$ and $\frac{4}{3}\pi$, $\cos 3\theta$ also equals 1, so the roots are 1, $\cos 2\pi/3 + i \sin 2\pi/3$, $\cos 4\pi/3 + i \sin 4\pi/3$, or $x_1 = 1$, $x_2 = (-1 + i\sqrt{3})/2$, and $x_3 = (-1 - i\sqrt{3})/2$. Why didn't we use the other angles, $\pi/3$, π, and $5\pi/3$? We could have, but notice that they make $r = -1$, so we shall come up with the same answers as before. (Check it out if you're not convinced. Also, cube the cube roots we found to make sure that the answer is 1.)

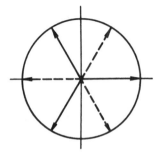

The roots, as you can see, are spread out evenly around a circle in the complex plane. Here is another way of arriving at the same result using only the complex exponential:

to solve $x^3 = 1$, let $x = re^{i\theta}$

then $r^3 e^{i3\theta} = 1$

The exponential provides a phase angle for a radius vector in the complex plane. This radius vector can lie on the real axis only if the phase angle is 0, or π, or $n\pi$. It will be *positive* and real only if $\theta = 0$ or *even* multiples of π. To satisfy this condition, θ can have values of 0, $2\pi/3$, $4\pi/3$, etc. When θ has these values, the exponential has the value of 1, and therefore r must equal 1. The cube roots of 1 are therefore $x = 1$, $e^{i(2/3)\pi}$, and $e^{i(4/3)\pi}$, which are the same as we derived using the sine and cosine notation.

11.3 APPLICATIONS OF COMPLEX NUMBERS

1. We claimed that complex numbers, in particular $e^{i\theta}$, were useful in describing physical phenomena. One powerful application is the description of wave motion.

$$y = A \sin 2\pi\left(\frac{t}{T} - \frac{x}{\lambda}\right) \tag{11-6}$$

Usually we think of the sine and cosine as being the sinusoidal functions, the ones describing the most common type of oscillations. The imaginary exponential is also sinusoidal. A very useful way of describing a traveling wave is

$$y = Ae^{i2\pi[(t/T)-(x/\lambda)]} \tag{11-7}$$

To be sure, in terms of sines and cosines the imaginary exponential is

$$y = Ae^{i2\pi[(t/T)-(x/\lambda)]}$$
$$= A\left[\cos 2\pi\left(\frac{t}{T} - \frac{x}{\lambda}\right) + i \sin 2\pi\left(\frac{t}{T} - \frac{x}{\lambda}\right)\right] \tag{11-8}$$

We get two of the ordinary kind of sinusoidal terms, one of them imaginary. They do not interfere with each other, however, at least under all the common conditions of linear equations where wave amplitudes can be superimposed. If the computations are simpler with the exponential form, and they often are, the exponential can be used throughout. At the end of the calculation, either the sine or the cosine part of the equivalent expression can be taken as the solution, since the two have not mixed with each other.

Here is a simple example. The differential equation for simple harmonic motion is

$$\frac{d^2y}{dt^2} = -\omega^2 y \tag{11-9}$$

This equation is satisfied for $y = A \sin \omega t$ or $y = A \cos \omega t$. It is also satisfied for $y = Ae^{\pm i \omega t}$. (Check this for yourself.) In this case there is no difficulty in working directly with the sine or cosine. When there is a friction term, however, the solution is easier to write with an exponential function.

$$\frac{d^2 y}{dt^2} + k\frac{dy}{dt} + \omega^2 y = 0 \tag{11-10}$$

Here the friction term, $k(dy/dt)$, is proportional to the velocity dy/dt. Let $y = Ae^{\alpha t}$. Then

$$\frac{dy}{dt} = \alpha A e^{\alpha t} \qquad \text{and} \qquad \frac{d^2 y}{dt^2} = \alpha^2 A e^{\alpha t} \tag{11-11}$$

Substituting these values into the equation leads to

$$\alpha^2 A e^{\alpha t} + k\alpha A e^{\alpha t} + \omega^2 A e^{\alpha t} = 0$$
$$\alpha^2 + k\alpha + \omega^2 = 0 \tag{11-12}$$

This is a quadratic in α with solutions

$$\alpha = \frac{-k \pm \sqrt{k^2 - 4\omega^2}}{2} \tag{11-13}$$

The solution to the differential equation is

$$y = Ae^{[(-k \pm \sqrt{k^2 - 4\omega^2})/2]t} = Ae^{(-k/2)t}e^{(\pm \sqrt{k^2 - 4\omega^2}/2)i} \tag{11-14}$$

The first exponential is real and represents a decay of the function. The second exponential can be either real or imaginary, depending on the relative sizes of the decay constant and the frequency. If $k^2 < 4\omega^2$, the term is imaginary and the exponential is sinusoidal. The solution to the differential equation could then be written as

$$y = Ae^{-(k/2)t}e^{[(\pm i\sqrt{4\omega^2 - k^2})/2]t}$$
$$= Ae^{-kt}\left[\sin \frac{\sqrt{4\omega^2 - k^2}}{2}t + i\cos \frac{\sqrt{4\omega^2 - k^2}}{2}t\right]$$

or

$$= Ae^{-kt}\left[-\sin \frac{\sqrt{4\omega^2 - k^2}}{2}t + i\cos \frac{\sqrt{4\omega^2 - k^2}}{2}t\right] \tag{11-15}$$

Since sine and cosine differ only by 90°; either can be chosen if an arbitrary phase angle is included. That phase angle and the amplitude are the two constants of integration, and are determined by the initial conditions:

$$y = Ae^{-kt} \sin\left(\frac{\sqrt{4\omega^2 - k^2}}{2}t + \phi\right) \tag{11-16}$$

2. An electromagnetic plane wave traveling in the x direction with velocity c can be described as

$$E = E_0 e^{i\omega[t - (x/c)]} \tag{11-17}$$

If the wave strikes a metal, some energy will be reflected and some will penetrate. An index of refraction can be assigned to the metal, based on a model of free charges that can have density oscillations with a natural frequency ω_p, called the *plasma frequency*. The conductivity of the metal is σ and the dielectric constant is ϵ. In terms of these parameters, the index of refraction is given by

$$n^2 = 1 - \frac{\omega_p^2}{\omega^2 - i[(\omega_p^2/\sigma)\epsilon]\omega} \tag{11-18}$$

When the frequency of the incoming electromagnetic wave is much less than ω_p, the plasma frequency, $n^2 \approx -i(\sigma/\epsilon\omega)$. Solving for n gives

$$n \approx \sqrt{\frac{\sigma}{\epsilon\omega}}\left(\frac{1-i}{\sqrt{2}}\right)$$

Substituting this into the expression for a plane wave,

$$E = E_0 e^{i\omega[t-(nx/c)]} = E_0 e^{-\sqrt{\sigma\omega/2\epsilon}(x/c)}e^{i\omega[t-\sqrt{\sigma/2\epsilon\omega}(x/c)]} \tag{11-19}$$

The term in front, $E_0 e^{-\sqrt{\sigma\omega/2\epsilon}(x/c)}$ is the amplitude of the refracted wave. The amplitude goes down exponentially in the metal. For the complex index of refraction and very low frequencies, the metal absorbs radiation; it isn't transparent. A graph showing such exponential decay is given in the diagram.

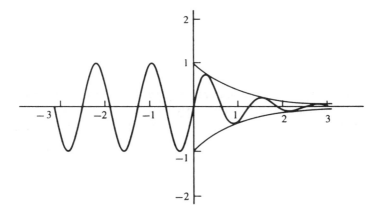

For very high frequencies, $\omega \gg \omega_p$, $n^2 \approx 1 - (\omega_p^2/\omega^2)$. This is a real number, less than one. $n \approx \sqrt{1 - (\omega_p^2/\omega^2)}$. The electric field in the metal then becomes $E = E_0 e^{i\omega[t-\sqrt{1-(\omega_p/\omega)^2}x/c]}$ and there is no absorption. Consequently, in agreement with experiment, metal is transparent to x rays and in some cases even to ultraviolet. One more point: for both cases the coefficient in front of the space coordinate is less than one. This corresponds to an increased wavelength in the metal.

11.3 APPLICATIONS OF COMPLEX NUMBERS

Another widespread use of complex numbers is in the description of alternating current. In ordinary dc circuits, the current is proportional to the applied voltage: $I = V/R$. The resistance of a circuit element determines the magnitude of current for a particular potential difference. With alternating current, circuit elements such as inductances and capacitors can determine not only the amplitude, but also the relative phases of current and voltage. In a "pure" capacitance (without resistance), the voltage lags the alternating current by 90°. In a pure inductance the voltage leads the current by 90°. In a series circuit with the same current (and at the same phase) in each component, the voltage across the inductance will be 180° ahead of the voltage across the capacitor. Thus it is possible to have large potential differences across the inductance and capacitor and yet have much smaller potential difference across the series combination. The out-of-phase potential differences cancel.

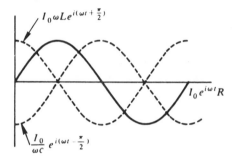

Complex numbers provide a natural way to describe this combination of magnitude and phase angle, a discovery that was first exploited by the great theoretical electrical engineer, Charles Steinmetz. The "reactance" of a capacitor is defined to be

$$X_c = -i\frac{1}{\omega C} \tag{11-20}$$

The reactance of an inductance is

$$X_L = i\omega L \tag{11-21}$$

The angular frequency, ω, equals $2\pi f$. With C in *farads* and L in *henries*, the reactances are in ohms. For a current, I, in amperes, the potential differences in volts across resistance, capacitor, and inductance are given by IR, $-i(I/\omega C)$, and $iI\omega L$. These can be visualized with a diagram in the complex plane.

COMPLEX NUMBERS

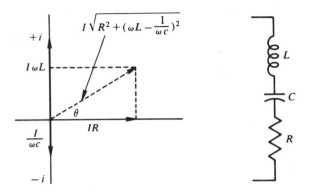

The voltage across the series circuit is the complex sum of the separate voltages.

$$V = I\left[R + i\left(\omega L - \frac{1}{\omega C}\right)\right] \qquad (11\text{-}22)$$

The magnitude of the voltage is the magnitude of the radius vector in the complex plane:

$$|V| = I\sqrt{R^2 + \left(\omega L - \frac{1}{\omega C}\right)^2} \qquad (11\text{-}23)$$

The phase angle is

$$\theta = \tan^{-1}\frac{\omega L - (1/\omega C)}{R} \qquad (11\text{-}24)$$

At resonance, $\omega L = (1/\omega C)$, the voltage is a minimum. For a given impressed voltage, such as in an antenna circuit, this condition yields maximum current and the current is in phase with the voltage.

The combination of resistance and reactance is called "impedance." The magnitude of the impedance is simply

$$|Z| = \sqrt{R^2 + \left(\omega L - \frac{1}{\omega C}\right)^2} \qquad (11\text{-}25)$$

Because current and voltage are 90° out of phase in inductances and capacitors, the power expended in them over a full period is zero. During half of each period, energy is being stored in an inductance or capacitor, and during the other half it is fed back to the system. Only the resistive components of the circuit absorb energy, turning it into heat. (We neglect energy that is radiated, which is small at low frequencies.)

The instantaneous product of current and voltage in a series circuit with L, C, and R is

$$IV = I_0^2 e^{2i\omega t}\sqrt{R^2 + \left(\omega L - \frac{1}{\omega C}\right)^2}\cos\theta \qquad (11\text{-}26)$$

From the geometry of the complex plane (see the diagram on p. 169),

(see the diagram on p. 169)

$$\cos \theta = \frac{R}{\sqrt{R^2 + [\omega L - (1/\omega C)]^2}} \qquad (11\text{--}27)$$

The average power absorbed in one complete cycle is

$$\frac{1}{T} \int_0^T IV \, dt = \tfrac{1}{2} I_0^2 R \qquad (11\text{--}28)$$

SAMPLES

1. Using the fact that $e^{i\theta} = \cos \theta + i \sin \theta$, express $\cos \theta$ and $\sin \theta$ in terms of the exponential function. $[\cos \theta = (e^{i\theta} + e^{-i\theta})/2, \; \sin \theta = (e^{i\theta} - e^{-i\theta})/2i]$

2. Perform the following operations:

   (a) $(1 + i)(3 + 4i)$ $\qquad\qquad\qquad\qquad\qquad$ $(-1 + 7i)$

   (b) $(2 - 2i)(5i)$ $\qquad\qquad\qquad\qquad\qquad\quad$ $(10 + 10i)$

   (c) $(3 + i)/(5 - 12i)$ $\qquad\qquad\qquad\qquad\quad$ $[(3 + 41i)/169]$

   (d) $(-5 + 2i)/(1 + 6i)$ $\qquad\qquad\qquad\qquad$ $[(7 + 32i)/37]$

   (e) $(\sqrt{3} + i)/2)^{(2+i)}$ $\qquad\qquad\qquad\qquad$ $[e^{-\pi \, 6}(1 + \sqrt{3} \, i)/2]$

   (f) $[(1 + i)/\sqrt{2}]^{(6-4i)}$ $\qquad\qquad\qquad\qquad$ $(-ie^{\pi})$

   (g) $(\cos \theta + i \sin \theta)^n$, Demoivre's theorem $\quad$ $(\cos n\theta + i \sin n\theta)$

3. Find all four roots of the equation $x^4 = -8 + 8\sqrt{3} \, i$. $(\sqrt{3} + i, -\sqrt{3} - i, -1 + \sqrt{3} \, i, 1 - \sqrt{3} \, i)$

4. Find the roots of the equation $x^2 - i = 0$. $[(1 + i)/2, -(1 + i)/2]$

CHAPTER12 CALCULUS—DIFFERENTIATION

12.1 DEFINITIONS. Calculus is the mathematics of change. In the section on functions we emphasized the importance of the slope of the graph of each function. This slope, which is the tangent line to a point on the graph, determines the change of the function as the independent variable changes. The slope is defined algebraically as $\Delta y/\Delta x$. Since this ratio may depend on the size of the base interval, Δx, a strict definition requires the slope to be equal to the limit of the ratio as Δx goes to zero. Slope $= \lim_{\Delta x \to 0}(\Delta y/\Delta x)$. For most of the points of functions dealt with in introductory science, such limits do exist. In the case illustrated, the limiting process produces a slope of 2.00 at $x = 1$, $y = 1$.

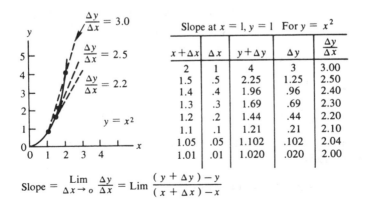

Slope at $x = 1$, $y = 1$ For $y = x^2$

$x + \Delta x$	Δx	$y + \Delta y$	Δy	$\frac{\Delta y}{\Delta x}$
2	1	4	3	3.00
1.5	.5	2.25	1.25	2.50
1.4	.4	1.96	.96	2.40
1.3	.3	1.69	.69	2.30
1.2	.2	1.44	.44	2.20
1.1	.1	1.21	.21	2.10
1.05	.05	1.102	.102	2.04
1.01	.01	1.020	.020	2.00

$$\text{Slope} = \frac{\text{Lim}}{\Delta x \to 0} \frac{\Delta y}{\Delta x} = \text{Lim} \frac{(y + \Delta y) - y}{(x + \Delta x) - x}$$

The functional form of the slope can be found analytically for most functions. In this form it is called the *derivative:*

$$\frac{dy}{dx} = \lim_{\Delta x \to 0} \frac{f(x + \Delta x) - f(x)}{\Delta x} \qquad \text{where } y = f(x) \qquad (12\text{--}1)$$

The dy and dx can be treated as finite terms if the limiting process works. They are called *differentials*.

If $y = f(x)$, it is frequently convenient to write $dy/dx = f'(x)$. Then $dy = f'(x)dx$. The differential of a function is equal to the product of its derivative and the differential of the independent variable.

12.2 DERIVATION OF DERIVATIVES. Here are derivations for the derivatives of three different functions.

1. $y = f(x) = x^2$.

$$\frac{dy}{dx} = \lim_{\Delta x \to 0} \frac{f(x + \Delta x) - f(x)}{\Delta x} = \frac{(x + \Delta x)^2 - x^2}{\Delta x}$$

$$= \frac{x^2 + 2x\,\Delta x + \Delta x^2 - x^2}{\Delta x} = 2x + \Delta x$$

As $\Delta x \to 0$,

$$\frac{dy}{dx} = 2x \qquad (12\text{--}2)$$

(In agreement with the graph and data of our earlier example, at $x = 1$, the derivative or slope $= 2$.)

2. $y = f(x) = \sin x$.

$$\frac{dy}{dx} = \lim_{\Delta x \to 0} \frac{f(x + \Delta x) - f(x)}{\Delta x} = \frac{\sin (x + \Delta x) - \sin x}{\Delta x}$$

$$= \frac{2 \cos (x + \tfrac{1}{2}\Delta x) \sin (\tfrac{1}{2}\Delta x)}{\Delta x}$$

[The last step is justified by the trig identity:

$$\sin \theta - \sin \varphi = 2 \cos \tfrac{1}{2}(\theta + \varphi) \sin \tfrac{1}{2}(\theta - \varphi)$$

Let $\theta = x + \Delta x$ and $\varphi = x$.]

$$= \cos (x + \tfrac{1}{2}\Delta x) \frac{\sin (\tfrac{1}{2}\Delta x)}{\tfrac{1}{2}\Delta x}$$

As $\Delta x \to 0$,

$$\frac{\sin (\tfrac{1}{2}\Delta x)}{\tfrac{1}{2}\Delta x} \to 1 \qquad \text{and} \qquad \cos (x + \tfrac{1}{2}\Delta x) \to \cos x$$

[The series expansions for $\sin (\tfrac{1}{2}\Delta x)$ and $\cos (x + \tfrac{1}{2}\Delta x)$ are (see p. 195)

$$\sin (\tfrac{1}{2}\Delta x) = (\tfrac{1}{2}\Delta x) - \frac{(\tfrac{1}{2}\Delta x)^3}{3!} + \cdots$$

$$\cos (x + \tfrac{1}{2}\Delta x) = 1 - \frac{(x + \tfrac{1}{2}\Delta x)^2}{2!} + \cdots$$

$$= 1 - \frac{x^2}{2!} + \cdots - \frac{x \, \Delta x + \frac{1}{4}\Delta x^2}{2!} + \cdots .]$$

Hence

$$\frac{d \sin x}{dx} = \cos x \qquad (12\text{--}3)$$

3. $y = u(x) \cdot v(x)$, the product of two functions.

$$(y + \Delta y) - y = (u + \Delta u)(v + \Delta v) - uv = uv + v\Delta u$$
$$+ u \, \Delta v + \Delta u \, \Delta v - uv$$
$$= u \, \Delta v + v \, \Delta u + \Delta u \, \Delta v$$
$$\frac{\Delta y}{\Delta x} = \frac{u \, \Delta v + v \, \Delta u + \Delta u \, \Delta v}{\Delta x} = u\frac{\Delta v}{\Delta x} + (v + \Delta v)\frac{\Delta u}{\Delta x}$$
$$\frac{dy}{dx} = u\frac{dv}{dx} + v\frac{du}{dx} \qquad (12\text{--}4)$$

The differential dx can be treated as a finite quantity, turning this equation into $dy = u \, dv + v \, du$.

We presented an example of this formula in another guise on p. 9 in the section on error analysis. The percentage (or fractional) uncertainty of a product of two quantities is equal to the sum of the percentage uncertainties of the individual quantities; e.g., $A = LW$, the area of a rectangle with length L and width W.

$$dA = L \, dW + W \, dL$$

Divide by the area A:

$$\frac{dA}{A} = \frac{L}{A}dW + \frac{W}{A}dL$$
$$\frac{dA}{A} = \frac{dW}{W} + \frac{dL}{L}$$

12.3 DERIVATIVES OF COMMON FUNCTIONS. Here is a table of derivatives of common functions. The derivations of these follow the same procedures as the three examples above, though each may involve some special algebraic trick.

$y = k$	$\dfrac{dy}{dx} = 0$
$y = x$	$\dfrac{dy}{dx} = 1$
$y = kx$	$\dfrac{dy}{dx} = k$
$y = u(x)v(x)$	$\dfrac{dy}{dx} = u\dfrac{dv}{dx} + v\dfrac{du}{dx}$
$y = x^n$	$\dfrac{dy}{dx} = nx^{n-1}$ (for any n, positive or negative)

173

$$y = \sin x \qquad \frac{dy}{dx} = \cos x$$

$$y = \cos x \qquad \frac{dy}{dx} = -\sin x$$

$$y = \tan x \qquad \frac{dy}{dx} = \sec^2 x$$

$$y = \cot x \qquad \frac{dy}{dx} = -\csc^2 x$$

$$y = \sec x \qquad \frac{dy}{dx} = \sec x \tan x$$

$$y = \csc x \qquad \frac{dy}{dx} = -\csc x \cot x$$

$$y = \sin^{-1} x \qquad \frac{dy}{dx} = (1 - x^2)^{-1/2}$$

$$y = \cos^{-1} x \qquad \frac{dy}{dx} = -(1 - x^2)^{-1/2}$$

$$y = \tan^{-1} x \qquad \frac{dy}{dx} = (1 + x^2)^{-1}$$

$$y = \cot^{-1} x \qquad \frac{dy}{dx} = -(1 + x^2)^{-1}$$

$$y = \sec^{-1} x \qquad \frac{dy}{dx} = x^{-1}(x^2 - 1)^{-1/2}$$

$$y = \csc^{-1} x \qquad \frac{dy}{dx} = -x^{-1}(x^2 - 1)^{-1/2}$$

$$y = \log_a x \qquad \frac{dy}{dx} = x^{-1} \log_a e$$

$$y = \ln x \qquad \frac{dy}{dx} = x^{-1}$$

$$y = a^x \qquad \frac{dy}{dx} = a^x \ln a$$

$$y = e^x \qquad \frac{dy}{dx} = e^x$$

12.4 CHAIN RULE OF DIFFERENTIATION. Frequently it is necessary to find the derivative of some function that is in turn a function of x. For instance, suppose that $f(x) = \sin^2 x = u^2$, where $u = \sin x$. The *chain rule for differentiation* is

$$\frac{df(x)}{dx} = \frac{df(x)}{du} \frac{du}{dx} \qquad\qquad (12\text{--}5)$$

In this case,

$$\frac{d \sin^2 x}{dx} = \frac{d(\sin^2 x)}{d(\sin x)} \frac{d(\sin x)}{dx} = 2 \sin x \cos x$$

Suppose that the *argument* of the function is not the independent variable, but is instead some function of it. For example,

$$\frac{d\sin^2\omega t}{dt} = \frac{d(\sin^2\omega t)}{d(\sin\omega t)}\frac{d(\sin\omega t)}{d(\omega t)}\frac{d(\omega t)}{dt}$$

$$= 2\sin\omega t\cdot\cos\omega t\cdot\omega = 2\omega\sin\omega t\cos\omega t$$

Here are some other examples of the chain rule.

1. $y = 1/\sqrt{x^2 + k} = (x^2 + k)^{-1/2}.$

$$\frac{dy}{dx} = \frac{dy}{du}\frac{du}{dx} = \frac{d(x^2 + k)^{-1/2}}{d(x^2 + k)}\frac{d(x^2 + k)}{dx}$$

$$= -\tfrac{1}{2}(x^2 + k)^{-3/2}2x = -\frac{x}{(x^2 + k)^{3/2}} \qquad (12\text{--}6)$$

2. $y = \dfrac{u(x)}{v(x)} = u(x)v^{-1}(x).$ (This is just a special case of a product of two functions.)

$$\frac{dy}{dx} = u\frac{dv^{-1}}{dx} + v^{-1}\frac{du}{dx}$$

$$= u\frac{d(v^{-1})}{d(v)}\frac{dv}{dx} + v^{-1}\frac{du}{dx} = -uv^{-2}\frac{dv}{dx} + v^{-1}\frac{du}{dx}$$

$$= \frac{v(du/dx) - u(dv/dx)}{v^2} \qquad (12\text{--}7)$$

3. $y = Ae^{-i\sqrt{\omega^2 t^2 - k^2}}.$

$$\frac{dy}{dt} = \dot{y} = \frac{Ad(e^{-i\sqrt{\omega^2 t^2 - k^2}})}{(-i\sqrt{\omega^2 t^2 - k^2})}\frac{d(-i\sqrt{\omega^2 t^2 - k^2})}{d(\omega^2 t^2 - k^2)}\frac{d(\omega^2 t^2 - k^2)}{dt}$$

$$= Ae^{-i\sqrt{\omega^2 t^2 - k^2}}[-\tfrac{1}{2}i(\omega^2 t^2 - k^2)^{-1/2}][2\omega^2 t] \qquad (12\text{--}8)$$

Note here the common symbol for differention by time:

$$\frac{dy}{dt} = \dot{y}$$

12.5 THE SECOND DERIVATIVE. The first derivative of a function corresponds graphically to its slope. If the derivative of the first derivative is taken, we find out how the *slope* is changing value. This second derivative is written as

$$\frac{d^2y}{dx^2}$$

Here are some second derivatives of common functions:

1. $y = kx^2 \qquad \dfrac{dy}{dx} = 2kx \qquad \dfrac{d^2y}{dx^2} = 2k \qquad (12\text{--}9)$

The function describes a parabola. The slope is proportional to x, and *its* slope is constant.

2. $y = \sin \omega t \qquad \dot{y} = \omega \cos \omega t \qquad \ddot{y} = -\omega^2 \sin \omega t$ $\qquad$ (12–10)

Note that the second derivative with respect to time is indicated by a double dot over the function. Here the second derivative is sinusoidal, as is the velocity and the original function. The second derivative is 180° out of phase with the function.

3. The first derivative with respect to time of a displacement is a velocity; the second derivative is an acceleration. $(dx/dt) = v$; $(d^2x/dt^2) = a$. The displacement of an object falling freely from rest is $y = \frac{1}{2} gt^2$. The velocity is $\dot{y} = gt$. The acceleration is $\ddot{y} = g$.

12.6 GEOMETRICAL APPLICATION OF DERIVATIVES. Wherever a smooth function has a minimum or maximum, its first derivative must equal zero. (The tangent at the point of maximum or minimum is horizontal; the slope is zero.) We can find these points of maximum or minimum by

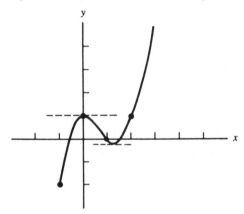

differentiating the function and setting this first derivative equal to zero. In the case illustrated, $y = x^3 - 2x^2 + 1$,

$$\frac{dy}{dx} = 3x^2 - 4x$$

Where $dy/dx = 0$, $y(x)$ is a maximum or minimum: $3x^2 - 4x = 0$. The equation is satisfied for $x = 0$ and $x = \frac{4}{3}$.

Now consider the second derivative of the function at the point of local maximum ($x = 0$) and local minimum ($x = \frac{4}{3}$).

$$\frac{d^2y}{dx^2} = 6x - 4 \qquad \left(\frac{d^2y}{dx^2}\right)_{x=0} = -4 \qquad \left(\frac{d^2y}{dx^2}\right)_{x=4/3} = +4$$

Notice that at the local maximum, the second derivative is negative. The

CALCULUS—DIFFERENTIATION

first derivative is zero, but the *slope* of the first derivative is negative; the function is decreasing for larger x. At the local minimum, the second derivative is positive. The slope of the first derivative is positive; the function is increasing for larger x. The polarity of the second derivative can be used as a test of whether the point with zero first derivative is a maximum or minimum.

A point of inflection is where the second derivative is equal to zero and the tangent at the point crosses the curve. For example, consider the function

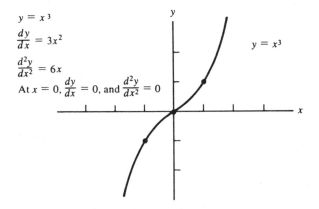

$y = x^3$

$\frac{dy}{dx} = 3x^2$

$\frac{d^2y}{dx^2} = 6x$

At $x = 0$, $\frac{dy}{dx} = 0$, and $\frac{d^2y}{dx^2} = 0$

$y = x^3$

At $x = 0$, $dy/dx = 0$, and $d^2y/dx^2 = 0$. The fact that the first derivative is zero does not guarantee a minimum or maximum.

12.7 CURVATURE OF A FUNCTION. The differential arc length is given by $ds^2 = dx^2 + dy^2$.

$$\frac{dy}{ds} = \sin \phi \qquad \frac{dx}{ds} = \cos \phi \qquad \frac{dy}{dx} = \tan \phi$$

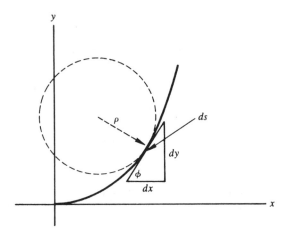

The *curvature* is defined as

$$\frac{d\phi}{ds} \qquad (12\text{--}11)$$

If there is no curvature, the angle ϕ is constant.

The *radius* of curvature is defined as the reciprocal of the curvature:

$$\rho = \frac{ds}{d\phi} \qquad (11\text{--}12)$$

It is the radius of a circle that would match the curvature of the function at that particular point. To express $\rho = ds/d\phi$ in terms of x and y, we substitute for ds and $d\phi$.

$$ds = \sqrt{dx^2 + dy^2}$$

$$\phi = \tan^{-1}\frac{dy}{dx} \qquad d\phi = \frac{1}{1 + (dy/dx)^2}d\left(\frac{dy}{dx}\right) = \frac{(d^2y/dx^2)dx}{1 + (dy/dx)^2}$$

Therefore,

$$\rho = \frac{(dx^2 + dy^2)^{1/2}}{(d^2y/dx^2)dx}\left[1 + \left(\frac{dy}{dx}\right)^2\right] = \frac{[1 + (dy/dx)^2]^{1/2}[1 + (dy/dx)^2]}{d^2y/dx^2}$$

$$\rho = \frac{[1 + (dy/dx)^2]^{3/2}}{d^2y/dx^2} \qquad (12\text{--}13)$$

Take the positive value for the radical in the numerator and ρ will have the sign of the second derivative. The radius of curvature is positive if the curve is concave upward, negative if the curve is concave downward.

Let us use these expressions to find the radius of curvature of the curve given in polar coordinates by $r = R\cos\theta$. In rectangular coordinates this is

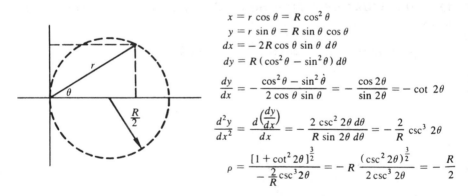

$$x = r\cos\theta = R\cos^2\theta$$
$$y = r\sin\theta = R\sin\theta\cos\theta$$
$$dx = -2R\cos\theta\sin\theta\, d\theta$$
$$dy = R(\cos^2\theta - \sin^2\theta)\, d\theta$$

$$\frac{dy}{dx} = -\frac{\cos^2\theta - \sin^2\theta}{2\cos\theta\sin\theta} = -\frac{\cos 2\theta}{\sin 2\theta} = -\cot 2\theta$$

$$\frac{d^2y}{dx^2} = \frac{d\left(\frac{dy}{dx}\right)}{dx} = -\frac{2\csc^2 2\theta\, d\theta}{R\sin 2\theta\, d\theta} = -\frac{2}{R}\csc^3 2\theta$$

$$\rho = \frac{[1 + \cot^2 2\theta]^{\frac{3}{2}}}{-\frac{2}{R}\csc^3 2\theta} = -R\frac{(\csc^2 2\theta)^{\frac{3}{2}}}{2\csc^3 2\theta} = -\frac{R}{2}$$

The curve has constant curvature. It is a circle as shown.

SAMPLES

1. What should be the radius and height of a cylindrical liter jar open at the top containing the least outside surface area?

CALCULUS—DIFFERENTIATION

The requirement is to minimize surface area A by appropriate choice of r with the restriction that $\pi r^2 h = V = 1$ liter.

$$A = \pi r^2 + 2\pi r h = \pi r^2 + \frac{2\pi r}{\pi r^2} V = \pi r^2 + \frac{2V}{r}$$

$$\frac{dA}{dr} = 2\pi r - \frac{2V}{r^2} = 0$$

$$r = \left(\frac{V}{\pi}\right)^{1/3} \qquad h = \left(\frac{V}{\pi}\right)^{1/3}$$

The radius should equal the height. Under these conditions the second derivative equals $d^2 A/dr^2 = 2\pi + 4(V/r^3)$. Since both r and V are positive, the second derivative is positive and the condition provides a minimum.

2. Find the launch angle that provides maximum range for a projectile that is not subject to air friction. Calculate R in terms of v_0 and θ. Differentiate

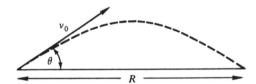

with respect to θ. Remember that $\cos^2 \theta - \sin^2 \theta = \cos 2\theta$. $(\theta = 45°.)$

3. For a given perimeter of a rectangle, what ratio of length to width yields maximum area? $(L = W$, the square.)

4. Find the derivative of x^3 by following the procedure of taking finite differences and then taking the limit of the ratio $\Delta y/\Delta x$.

5. Find the derivative of e^x in the same way as in (4). (Make use of the series approximation for e^x.)

CHAPTER13INTEGRATION

For every function useful in introductory science there is a derivative: $dF(x)/dx = f(x)$. We call $f(x)$ the derivative of $F(x)$, and could also define $F(x)$ as the antiderivative of $f(x)$. By itself, however, $F(x)$ is not unique. We could add any constant, so that $d(F(x) + C)/dx = f(x)$. The total antiderivative of $f(x)$ is thus $F(x) + C$. Another way of writing this is in terms of the symbol for the indefinite integral:

$$F(x) + C = \int f(x)\, dx \qquad (13\text{--}1)$$

Differentiating both sides gives $dF(x)/dx + 0 = f(x)$, since the derivative of an indefinite integral is just the function inside the integral sign.

There are no general rules for finding the antiderivatives, or integrals, of functions. The simple functions used most often, such as x^n, a^x, $\sin x$, etc., have simple derivatives and integrals that should be memorized. There are tricks to reduce more complicated integrals to simpler or standard forms. For practical purposes in introductory science, you can find the integrals you need in tables. We provide a short one in this book (see Appendix 10).

13.1 INTEGRALS OF SIMPLE FUNCTIONS. Here are some examples of integrals of the simpler functions.

Since $(d/dx)x^n = nx^{n-1}$, then $\int x^n\, dx = [1/(n + 1)]x^{n+1} + C$. (Except for $n = -1$, in which case $\int x^{-1}\, dx = \ln x + C$. See p. 66.)

Since $(d/dx)e^{kx} = ke^{kx}$, then $\int e^{kx}\, dx = (1/k)e^{kx} + C$.

Since $(d/dx)\sin kx = k \cos kx$, then $\int \cos kx\, dx = (1/k)\sin kx + C$.

13.2 INTEGRATION BETWEEN LIMITS. Integration between limits is basically a summing process. Plot the derivative $f(x)$ of some function $F(x)$, with $f(x)$ as the ordinate and x as the abscissa. The crosshatched area of the

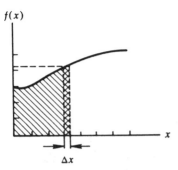

diagram has the approximate value $f(x)\,\Delta x$. [That is only an approximation to the area, since $f(x)$ changes from x to $x + \Delta x$.] The total shaded area under the curve, starting at $x = 0$, changes by $f(x)\,\Delta x$ as x increases by Δx. In the limit as $\Delta x \to 0$, the approximation gets better, and

$$dA = f(x)\,dx \qquad \text{or} \qquad \frac{dA}{dx} = f(x) \tag{13-2}$$

This is the expression relating a function and its derivative. By definition, $A = \int f(x)\,dx$, and the area under the curve is the same as the antiderivative $F(x)$. The complete expression must include a constant of integration that can be evaluated in terms of the area. The range of the variable x is shown on the integral sign by writing an upper and lower limit. If we want the area under the curve between $x = a$ and $x = b$,

$$A_{a-b} = F(x) + C = \int_a^b f(x)\,dx \tag{13-3}$$

The constant, C, must have a value so that if we took the integral from a to a, yielding no area, $F(a) + C = 0$. Therefore $C = -F(a)$. When $x = b$, $F(x) = F(b)$. The definite integral can thus be evaluated:

$$\int_a^b f(x)\,dx = F(x)\Big|_a^b = F(b) - F(a) \tag{13-4}$$

For instance, if $f(x) = kx$, $F(x) = \tfrac{1}{2}kx^2$. Then

$$\int_0^b f(x)\,dx = \int_0^b kx\,dx = F(x)\Big|_0^b = \tfrac{1}{2}kx^2\Big|_0^b = \tfrac{1}{2}kb^2 - 0$$

The shaded area of the diagram on p. 182 is a triangle with base b and height $f(b) = kb$. Its area is therefore $\tfrac{1}{2}kb^2$, which agrees with the integration.

In principle we could find any definite integral by plotting it and finding the area under the curve between the given limits. We would not have to know the antiderivative of the particular function. As an example, consider

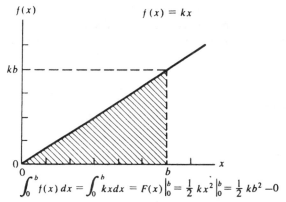

$f(x)$ $f(x) = kx$

kb

0

b

x

$$\int_0^b f(x)\,dx = \int_0^b kx\,dx = F(x)\Big|_0^b = \tfrac{1}{2}kx^2\Big|_0^b = \tfrac{1}{2}kb^2 - 0$$

a force that is a function of a displacement, x. The dot, or scalar, product of force and displacement is the work done by the force:

$$dW = F \cdot dx$$

If $F = K$, the total work done in a displacement is represented by the rectangular area in the graph. The summation can also be written $W =$

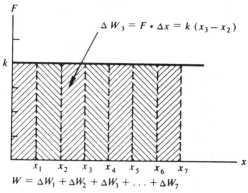

F

$\Delta W_3 = F \cdot \Delta x = k\,(x_3 - x_2)$

k

x_1 x_2 x_3 x_4 x_5 x_6 x_7 x

$$W = \Delta W_1 + \Delta W_2 + \Delta W_3 + \ldots + \Delta W_7$$

$\sum_{i=1}^7 \Delta W_i$. The capital sigma is a symbol implying the summation of all the ΔW_i, as i goes from 1 to 7. In this case, since $F = k$, the expression is exact and it makes no difference whether the Δx intervals are large or small.

Suppose that $F = kx$. The graph and work intervals now look as indicated in the diagram. The work done by the force in pushing from 0 to x_1 must be

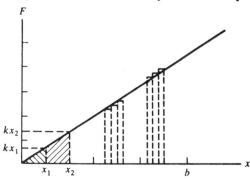

F

kx_2

kx_1

x_1 x_2 b x

given by the product of some average force and the displacement. The force is zero at the beginning of the interval, and kx_1 at the end. Even if we make the intervals smaller, it is necessary to take some sort of average force in the interval. In the first group of narrow intervals on the diagram, the height of each rectangle is taken to be kx. The sum of these rectangular areas is less than the total area. In the second group of narrow intervals, the height of each rectangle is taken to be $k(x + \Delta x)$. The sum of these rectangular areas is more than the total area. To eliminate the correction, we should let $\Delta x \to 0$. The summation, $\sum_{i=1}^{7}$ becomes an integral, $\int_0^b$.

$$W = \lim_{\Delta x \to 0} \sum_{i=1}^{n} kx_i \, \Delta x_i = \int_0^b kx \, dx \qquad (13\text{–}5)$$

We represent work graphically in terms of an "area," but of course this area does not have the L^2 dimensions of ordinary area. The dimensions of the graphical area are those of the product of ordinate and abscissa dimensions; in this case, (force·displacement) $\to (ML/T^2) \cdot L = M^1 L^2 T^{-2}$.

Frequently it is necessary to integrate by summing finite contributions. This is laborious but not difficult, particularly if care is taken to use small intervals when the function is changing rapidly and large intervals when the function remains relatively constant or can be approximated by an average. Computers routinely use numerical integration, rather than store large tables of integrals. For instance, if you ask a computer to integrate $\int_0^{90°} \sin \theta \, d\theta$, it may go into an internal subroutine that quickly sums $\sin \theta \, \Delta\theta$ from $\theta = 0$ to $\theta = 90°$, instead of referring to an internal library to find that $\int \sin \theta = -\cos \theta$.

13.3 METHODS OF INTEGRATION.

Let us find $\int_0^{90°} \sin \theta \, d\theta$ by three different methods:

1. *Integral tables.*

$$\int_0^{90°} \sin \theta \, d\theta = -\cos \theta \Big|_0^{90°} = -\cos 90° + \cos 0° = -0 + 1 = 1$$

2. *Numerical integration.*

$$A \approx \sum_{i=1}^{5} \Delta A_i = \frac{\pi}{6} \sin 15° + \frac{\pi}{9} \sin 40° + \frac{\pi}{18} \sin 55° + \frac{\pi}{18} \sin 65°$$

$$+ \frac{\pi}{9} \sin 80°$$

$$= 0.136 + 0.224 + 0.143 + 0.158 + 0.344$$

$$= 1.005$$

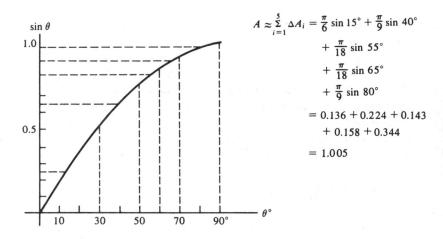

$$A \approx \sum_{i=1}^{5} \Delta A_i = \frac{\pi}{6} \sin 15° + \frac{\pi}{9} \sin 40°$$
$$+ \frac{\pi}{18} \sin 55°$$
$$+ \frac{\pi}{18} \sin 65°$$
$$+ \frac{\pi}{9} \sin 80°$$
$$= 0.136 + 0.224 + 0.143$$
$$+ 0.158 + 0.344$$
$$= 1.005$$

The difference between this answer and the first one may be due as much to the use of slide-rule calculations to three significant figures as it is to the averaging process. We took a large interval to begin with, since $\sin \theta$ is nearly linear from 0 to 30°. Assuming linearity in each interval, we used for the average $\sin ((\theta_f + \theta_i)/2)$. Someone else might have chosen different intervals and different averages. Usually, reasonable choices can be made just by inspecting the graph. As an example of an unreasonable choice, what if just one interval had been used with $\sin \theta$ at the midpoint, 45°?

$$A \approx \frac{\pi}{2} \sin 45° = 1.110$$

Notice that although we have cited θ in terms of degrees, the calculus formulas are always given with θ in terms of radians. An interval of 30° therefore gives an interval base of $\pi/6$.

3. Actual area measurement. If a function is carefully plotted on a sheet of graph paper, its integral between two limits can be found by counting up the area of the graph squares that are under the curve and between the limits.

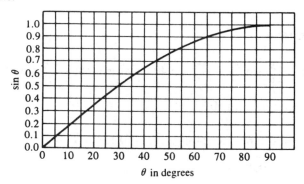

By actual count (try it yourself just once) there are 113.3 of the "quarter" squares (5 mm × 5 mm) under the curve from 0 to 90°. The number of such squares in the rectangle with base 0 to 90° and height 1.0 is 180. The area of that rectangle would have a value $= (\pi/2) \times 1.0 = 1.57$. Therefore, the area under the sine curve $= (113.3/180) \times 1.57 = 0.99$.

A similar but alternative way, known as chemist's integration, is to plot the function carefully on a graph paper, cut out the enclosed area, weigh it on an analytical balance, and then convert to the appropriate units by using the density of the graph paper.

13.4 SPECIAL TECHNIQUES FOR INTEGRATION. Although in practice you can look up integrals in tables instead of figuring them out, there are two special tricks that you should know. They are frequently used in derivations.

1. Integration by parts. The chain rule for differentiation is

$$d(uv) = u\,dv + v\,du$$

Integrating both sides,

$$uv = \int u\,dv + \int v\,du$$

There is a constant of integration that we can bring in later.

Changing the viewpoint to that of trying to integrate the function u, we have

$$\int u\,dv = uv - \int v\,du \qquad (13\text{--}6)$$

It might appear that not much has been gained, but sometimes the integral on the right is easier to solve than the one on the left. For example, $\int x \sin x\,dx$ is not one of the standard simple integrals. If we let $u = x$ and $dv = \sin x\,dx$, then $du = dx$ and $v = -\cos x$. Substituting into the formula,

$$\int x \sin x\,dx = -x \cos x + \int \cos x\,dx = -x \cos x + \sin x + C$$

Of course $\int x \sin x\,dx$ can also be found in most tables.

2. Change of variable. Even with an extensive reference table it may be necessary to convert your integral expression to match a standard form. Here is a trivial example. $\int e^x\,dx = e^x + C$. What is $\int e^{ax}\,dx$? In effect, we have $\int e^u\,dx$, where $u = ax$. It is generally true that $du = (du/dx)\,dx$. In this case, $du/dx = a$, and $du = a\,dx$. Changing the variable from x to u, our integral becomes $\int e^u(du/a) = 1/a \int e^u\,du = (1/a)e^u + C = (1/a)e^{ax} + C$.

If you change variables in a definite integral, then you must change limits also. For example: $\int_0^1 \sin \omega t\,dt$. The integral is to be taken between

$t = 0$ and $t = 1$, presumably measured in seconds. To get the integral into the standard form of $\int \sin \theta \, d\theta$, we let $\theta = \omega t = 2\pi f t = (2\pi/T)t$, and $d\theta = \omega \, dt$.

Then

$$\int_0^1 \sin \omega t \, dt = \int_{t=0}^{t=1} \sin \theta \frac{d\theta}{\omega} = \frac{1}{\omega} \int_0^{2\pi(1/T)} \sin \theta \, d\theta$$

The upper limit for the variable, θ, is now a fraction of 2π radians; $2\pi(1/T)$.

For a more complicated example of the usefulness of changing variables, see the derivation of the conic section orbits on p. 148. There a formula in r was reduced to a simpler and standard form by letting $u = 1/r$ and $du = -(1/r^2) \, dr$.

For a table of common integrals, see Appendix 6, p. 248.

13.5 APPLICATIONS OF INTEGRATION

A. Area of circle:

Polar coordinates.

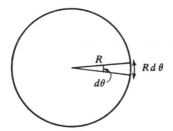

The area of the infinitesimal triangular slice is $dA = \frac{1}{2}RR \, d\theta$.

$$A = \int_0^{2\pi} \tfrac{1}{2} R^2 d\theta = \tfrac{1}{2} R^2 \theta \Big|_0^{2\pi} = \pi R^2$$

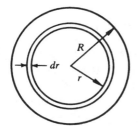

The area of the infinitesimal hoop is $dA = 2\pi r \, dr$.

$$A = \int_0^R 2\pi r \, dr = 2\pi \tfrac{1}{2} r^2 \Big|_0^R = \pi R^2$$

Cartesian coordinates.

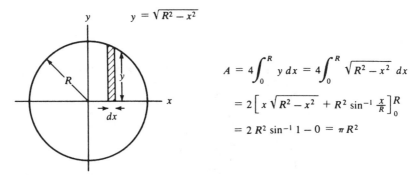

$$A = 4\int_0^R y\,dx = 4\int_0^R \sqrt{R^2 - x^2}\,dx$$

$$= 2\left[x\sqrt{R^2 - x^2} + R^2 \sin^{-1}\frac{x}{R}\right]_0^R$$

$$= 2R^2 \sin^{-1} 1 - 0 = \pi R^2$$

B. Line integrals: In one sense line integrals are just summations of elements of arc that yield a perimeter. They are no different from any other kind of integral, however, since they end up as summations of a "weighted" variable.

Circumference of circle—polar coordinates (the only sensible way):

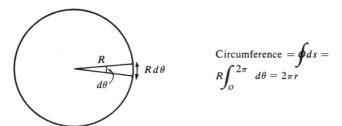

$$\text{Circumference} = \oint ds =$$

$$R\int_0^{2\pi} d\theta = 2\pi r$$

Circumference of circle—Cartesian coordinates (the hard way, to illustrate the "weighted" variable):

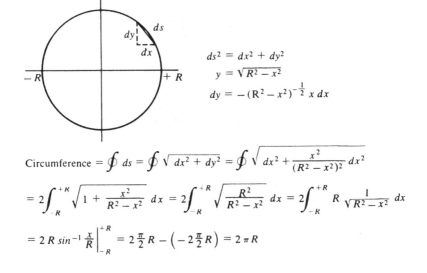

$$ds^2 = dx^2 + dy^2$$

$$y = \sqrt{R^2 - x^2}$$

$$dy = -(R^2 - x^2)^{-\frac{1}{2}} x\,dx$$

$$\text{Circumference} = \oint ds = \oint \sqrt{dx^2 + dy^2} = \oint \sqrt{dx^2 + \frac{x^2}{(R^2 - x^2)^2}\,dx^2}$$

$$= 2\int_{-R}^{+R} \sqrt{1 + \frac{x^2}{R^2 - x^2}}\,dx = 2\int_{-R}^{+R}\sqrt{\frac{R^2}{R^2 - x^2}}\,dx = 2\int_{-R}^{+R} R\,\frac{1}{\sqrt{R^2 - x^2}}\,dx$$

$$= 2R \sin^{-1}\frac{x}{R}\Big|_{-R}^{+R} = 2\frac{\pi}{2}R - \left(-2\frac{\pi}{2}R\right) = 2\pi R$$

C. Volume of a sphere—spherical coordinates:

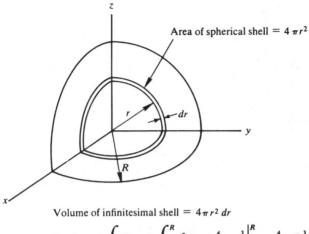

Area of spherical shell $= 4\pi r^2$

Volume of infinitesimal shell $= 4\pi r^2 \, dr$

$$V \text{ sphere} = \int dV = 4\pi \int_0^R r^2 \, dr = \tfrac{4}{3}\pi r^3 \Big|_0^R = \tfrac{4}{3}\pi R^3$$

D. Moment of inertia of sphere around z axis—cylindrical coordinates:

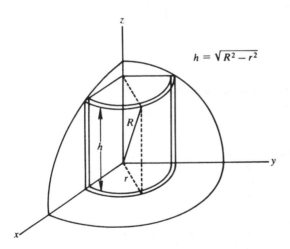

$$h = \sqrt{R^2 - r^2}$$

Moment of inertia of infinitesimal cylindrical shell around z axis

$$= \overbrace{2\pi r \sigma 2h \, dr}^{m} \; r^2$$

The mass density is σ. The half-height of the cylinder is h (only an upper octant of the sphere is shown).

$$I = 4\pi\sigma \int_0^R r^3 \sqrt{R^2 - r^2}\, dr \qquad (\text{Let } u = r^2,\ du = 2r\, dr)$$

$$= 2\pi\sigma \int_0^{\sqrt{U}} u\sqrt{R^2 - u}\, du = 2\pi\sigma\left[\frac{2(3u + 2R^2)}{15}\sqrt{(R^2 - u)^3}\right]_0^{\sqrt{U}}$$

$$= -2\pi\sigma\left[\frac{6r^2 + 4R^2}{15}\sqrt{(R^2 - r^2)^3}\right]_0^R = 2\pi\sigma\frac{4R^2}{15}R^3$$

$$= \tfrac{4}{3}\pi\sigma R^3 \tfrac{2}{5}R^2 = \tfrac{2}{5}MR^2$$

The mass of the sphere, M, is $\tfrac{4}{3}\pi\sigma R^3$. The change of variable in this case reduced the integral to a standard form.

SAMPLES

1. Find the moment of inertia of a solid disk with mass, M, and radius, R, rotating around its own axis. For an object at distance r from the axis and with mass m, the moment of inertia, I, is equal to mr^2. $(I = \tfrac{1}{2}MR^2.)$

2. Approximately how much work is done in cocking a crossbow with the force function shown in the graph? $(\approx 27\text{J}.)$

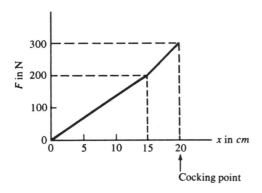

3. The Gaussian integral is described on p. 87 and p. 89 and there is a table of its values in Appendix 9. Evaluate the integrand at 6 points between $x = \bar{\mu}$ and $x = \bar{\mu} + 2\sigma$. Sketch the curve and from the graph find approximate values for the integral from $\bar{\mu}$ to $\bar{\mu} + \tfrac{1}{2}\sigma$, $\bar{\mu}$ to $\bar{\mu} + \tfrac{3}{2}\sigma$; and $\bar{\mu}$ to $\bar{\mu} + 2\sigma$.

4. The exact solution to simple pendulum motion leads to an elliptic integral for the period (see p. 208).

13.5 APPLICATIONS OF INTEGRATION

$$T = 2\sqrt{\frac{2}{k}} \int_0^{\theta_0} \frac{d\theta}{\sqrt{\cos \theta - \cos \theta_0}}$$

An elliptic integral has no exact solution. Obtain numerical values for the integral when $\theta_0 = 45°$. Compare your numerical integration with the series approximation given on p. 208.

CHAPTER14 SERIES AND

APPROXIMATIONS

In most applications of science, as well as in everyday life, we seldom need precision better than one or two significant figures. (One exception is in reconciling bank statements.) Usually we can use approximations to functions if it is awkward to handle the function itself. In this section we present some examples of how simple functions can be used as approximations to complex ones.

14.1 TAKING SQUARE ROOTS. Several complicated methods for finding square roots are taught in the schools, usually by rote. A logically satisfying way of obtaining a square root easily to two significant figures is illustrated in the following example.

Find the square root of 27. The answer is $5 + x$, where x is small compared with 5. Necessarily,

$$(5 + x)^2 = 27 \qquad \text{or} \qquad 25 + 10x + x^2 = 27$$

since $x^2 \ll 10x$, $10x \approx 2$, and $x \approx 0.2$. The square root of $27 \approx 5.2$.

Of course, the same method could be extended to find the next significant figure, but frequently that can be done more easily with a calculator. For illustration, however,

$$(5.2 + x)^2 = 27 \qquad \text{or} \qquad 27.04 + 10.4x + x^2 = 27$$
$$10.4x \approx -0.04 \qquad\qquad\qquad x \approx -0.004$$

Therefore,

$$\text{the square root of } 27 \approx 5.196$$

In this case we rapidly obtained high precision because our first guess was very close. Suppose that we had guessed that the square root of 27 is close to 6.

$$(6 + x)^2 = 27 \qquad 36 + 12x + x^2 = 27$$
$$12x \approx -9 \qquad x \approx -0.75$$

Our first approximation in this case is that $\sqrt{27} \approx 5.25$. Another example: What is the square root of 734?

$$\sqrt{734} = \sqrt{7.34} \times 10^1 = (3 + x) \times 10^1$$
$$(3 + x)^2 = 9 + 6x + x^2 = 7.34$$
$$6x \approx -1.66 \qquad x \approx -0.27$$

Therefore,

$$\sqrt{734} \approx 27.3$$

With a little practice, the procedure can be done in your head or with just a few steps noted on paper. Any of the school methods that require multiplication of intermediate numbers by 2 or 20 are equivalent to this method. The $\underline{2}$ arises from the coefficient of the middle term: $(a + x)^2 = a^2 + \underline{2}ax + x^2$.

14.2 THE BINOMIAL EXPANSION.

If n is a positive integer,

$$(a + b)^n = a^n + na^{n-1}b + \frac{n(n - 1)}{2!}a^{n-2}b^2 + \frac{n(n - 1)(n - 2)}{3!}a^{n-3}b^3$$

$$+ \cdots + \frac{n!}{k!(n - k)!}a^{n-k}b^k + \cdots + b^n \qquad (14\text{-}1)$$

For instance, $(a + b)^3 = a^3 + 3a^2b + 3ab^2 + b^3$.

The expansion also holds under restricted conditions when n is negative or a noninteger. In that case the expansion becomes an infinite series. For instance, here is the geometric series for $1/(1 - x) = (1 - x)^{-1}$.

$$\frac{1}{1 - x} = 1 + x + x^2 + x^3 + \cdots \qquad (14\text{-}2)$$

This result can also be obtained simply by dividing 1 by $1 - x$. In general, for negative or nonintegral values of n,

$$(1 + x)^n = 1 + nx + \frac{n(n - 1)}{2!}x^2 + \frac{n(n - 1)(n - 2)}{3!}x^3$$

$$+ \cdots + \frac{n!}{k!(n - k)!}x^k + \cdots \qquad (14\text{-}3)$$

This series converges and has a finite value if the following limit of ratios holds:

$$\lim_{k \to \infty} \left| \frac{a_{k+1}}{a_k} \right| < 1$$

where a_k is the kth term of the series.
The ratio is

$$\left|\frac{n!x^{k+1}}{(k+1)![n-(k+1)]!}\frac{k!(n-k)!}{n!x^k}\right| = \left|\frac{n-k}{k+1}x\right|$$

In the limit, as $k \to \infty$, this ratio goes to $|-x|$. The series converges if $|x| < 1$, or if $|x| = 1$ for $n > 0$.

As an example of the use of the binomial series to approximate a function, here is the series expansion for $\sqrt{1+x}$:

$$(1+x)^{1/2} = 1 + \frac{1}{2}x + \frac{\frac{1}{2}(-\frac{1}{2})}{2!}x^2 + \frac{\frac{1}{2}(-\frac{1}{2})(-\frac{3}{2})}{3!}x^3 + \cdots$$

$$= 1 + \frac{1}{2}x - \frac{1}{8}x^2 + \frac{1}{16}x^3 + \cdots \tag{14-4}$$

14.3 POWER SERIES.

A "power" series in x is defined to be

$$a_0 + a_1x + a_2x^2 + a_3x^3 + \cdots + a_nx^n + \cdots = \sum_{k=0}^{\infty} a_kx^k \tag{14-5}$$

If k runs only up to n, the series becomes a polynomial of order n.

The question of whether or not the infinite series converges (and if so, the range of x) depends on the values of the coefficients. There are a number of convergence tests, including the ratio test that we used for the binomial series.

Our problem is to find the appropriate coefficients for a given function. If a power series converges, its derivative exists for the range of x in which there is convergence. The differentiation can be carried out term by term. Let

$$f(x) = \sum_{k=0}^{\infty} a_kx^k = a_0 + a_1x + a_2x^2 + a_3x^3 + \cdots$$

Then

$$f'(x) = \sum_{k=1}^{\infty} ka_kx^{k-1} = a_1 + 2a_2x + 3a_3x^2 + 4a_4x^3 + \cdots$$

and

$$f''(x) = \sum_{k=2}^{\infty} k(k-1)a_kx^{k-2} = 2a_2 + 6a_3x + 12a_4x^2 + \cdots$$

in general

$$f^{(n)}(x) = \sum_{x=n}^{\infty} k(k-1)\cdots(k-n+1)a_kx^{k-n} \tag{14-6}$$

[Remember that the notation, $f'(x)$ is shorthand for $df(x)/dx$, and $f^{(n)}(x)$ symbolizes the nth derivative with respect to x.]

It can be shown that each of these series for the derivatives also converges for the same range of x for which $f(x)$ converges (except possibly at the end points). Evaluate each of the derivatives at $x = 0$:

$$f(0) = a_0$$
$$f'(0) = a_1$$
$$f''(0) = 2a_2$$
$$f'''(0) = 6a_3$$
$$f^{(n)}(0) = n!a_n$$

Now we know the coefficients of the original series.

$$f(x) = f(0) + f'(0)x + \frac{f''(0)}{2!}x^2 + \frac{f'''(0)}{3!}x^3 + \cdots$$
$$+ \frac{f^{(n)}(0)}{n!}x^n + \cdots \tag{14–7}$$

This is called Maclaurin's series and is useful for an expansion that approximates a function in the region around $x = 0$. The similar expansion in a region around $x = c$ is called the Taylor's series. It is

$$f(x) = f(c) + f'(c)(x - c) + \frac{f''(c)}{2!}(x - c)^2 + \cdots$$
$$+ \frac{f^{(n)}(c)}{n!}(x - c)^n + \cdots \tag{14–8}$$

As an example of the use of Maclaurin's series, let us find the expansion for $\sin x$ in the region near $x = 0$.

$$f(x) = \sin x \qquad f(0) = 0 \qquad f'''(x) = -\cos x \qquad f'''(0) = -1$$
$$f'(x) = \cos x \qquad f'(0) = 1 \qquad f''''(x) = \sin x \qquad f''''(0) = 0$$
$$f''(x) = -\sin x \qquad f''(0) = 0$$

$$\sin x = x - \frac{1}{3!}x^3 + \frac{1}{5!}x^5 - \frac{1}{7!}x^7 + \cdots \tag{14–9}$$

The series is good for every value of x. Graphs of several successive sums of terms are shown in the diagram on p. 264. Notice how $\sin x \approx x$ within $30°$ or so of $x = 0$. (x is measured in radians, of course.) All the terms of the series must be odd powers of x, since $\sin x$ is an odd function.

$$[\sin x = -\sin(-x)]$$

The similar expansions for $\cos x$ are shown in the diagram on p. 264. All the terms for $\cos x$ must be even powers of x, since $\cos x$ is an even function. [$\cos x = +\cos(-x)$].

Here is one more example of the use of Maclaurin's series:

$$f(x) = e^x \qquad f(0) = 1 \qquad f^{(n)}(x) = e^x \qquad f^{(n)}(0) = 1$$
$$f'(x) = e^x \qquad f'(0) = 1$$

It appears that all the derivatives are 1 at $x = 0$.

$$\therefore \quad e^x = 1 + x + \frac{x^2}{2!} + \frac{x^3}{3!} + \cdots \tag{14–10}$$

$$\sin x = x - \frac{x^3}{3!} + \frac{x^5}{5!} - \frac{x^7}{7!} + \cdots$$

$$\cos x = 1 - \frac{x^2}{2!} + \frac{x^4}{4!} - \frac{x^6}{6!} + \cdots$$

See Appendix 10 for graphs of the series expansions of $\sin x$ and $\cos x$.

$$\tan x = +\frac{x^3}{3} + \frac{2x^5}{15} + \frac{17x^7}{315} + \cdots \qquad (x^2 < 1)$$

$$= \frac{\pi}{2} - \frac{1}{x} + \frac{1}{3x^3} - \frac{1}{5x^5} + \cdots \qquad (x^2 > 1)$$

$$e^x = 1 + x + \frac{x^2}{2!} + \frac{x^3}{3!} + \cdots$$

$$a^x = 1 + x \ln a + \frac{(x \ln a)^2}{2!} + \frac{(x \ln a)^3}{3!} + \cdots$$

$$\sinh x = x + \frac{x^3}{3!} + \frac{x^5}{5!} + \frac{x^7}{7!} + \cdots$$

$$\cosh x = 1 + \frac{x^2}{2!} + \frac{x^4}{4!} + \frac{x^6}{6!} + \cdots$$

$$\ln(1 + x) = x - \frac{x^2}{2} + \frac{x^3}{3} - \frac{x^4}{4} + \frac{x^5}{5} \cdots \qquad (\text{for } |x| < 1)$$

$$\sin^{-1} x = x + \frac{1}{2}\frac{x^3}{3} + \frac{1}{2}\frac{3}{4}\frac{x^5}{5} + \frac{1}{2}\frac{3}{4}\frac{5}{6}\frac{x^7}{7} + \cdots$$

$$\tan^{-1} x = x - \frac{1}{3}x^3 + \frac{1}{5}x^5 - \frac{1}{7}x^7 + \cdots \qquad (x^2 < 1)$$

$$\ln x = 2\left[\frac{x-1}{x+1} + \frac{1}{3}\left(\frac{x-1}{x+1}\right)^3 + \frac{1}{5}\left(\frac{x-1}{x+1}\right)^5 + \cdots\right] \qquad x > 0$$

Geometric progression

$$\sum_{i=0}^{n} ar^i = a(1 - r^{n-1})/(1 - r) \qquad \text{For } r^2 < 1 \quad \sum_{i=0}^{\infty} ar^i = a/(1 - r)$$

14.4 FOURIER SERIES. Although a power series, if enough terms are used, can represent a periodic function such as a sine or cosine, it makes more sense to approximate periodic functions with a series of terms that are also periodic. This can be done by adding together sine and cosine terms of various frequencies. The physical analogy is the representation of a musical note by its fundamental and harmonics. The sounds from a flute, trumpet, and violin can be distinguished and identified even though they are all the same note; i.e., they all have the same fundamental. The flute produces a "purer" tone than the violin; it has fewer harmonics.

Here are three wave forms that have the same basic pitch. The first is a pure sine wave with no harmonics. The second has a strong third harmonic at three times the main frequency. (In this nomenclature, the first harmonic is the fundamental.) The third note is evidently rich in harmonics of high number, since there are so many small, sharp wiggles in the graph. The general shape, however, is domination by a second harmonic with twice the frequency of the fundamental.

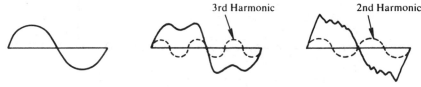

3rd Harmonic   2nd Harmonic

Any finite continuous function that is periodic in 2π can be represented by

$$f(x) = \tfrac{1}{2}a_0 + a_1 \cos x + a_2 \cos 2x + \cdots + b_1 \sin x$$
$$+ b_2 \sin 2x + \cdots$$
$$= \tfrac{1}{2}a_0 + \sum_{n=1}^{\infty} (a_n \cos nx + b_n \sin nx) \tag{14-11}$$

This is called a Fourier series. Actually, the representation is valid even if there are some breaks or step characteristics to the function. The series gives an average value at the step.

If the period is T rather than 2π, the series transforms to

$$f(t) = \frac{1}{2}a_0 + \sum_{n=1}^{\infty} \left(a_n \cos \frac{2\pi}{T} nt + b_n \sin \frac{2\pi}{T} nt \right) \tag{14-12}$$

If the period is spacelike, such as a wavelength, λ, the series is

$$f(x) = \frac{1}{2}a_0 + \sum_{n=1}^{\infty} \left(a_n \cos \frac{2\pi}{\lambda} nx + b_n \sin \frac{2\pi}{\lambda} nx \right) \tag{14-13}$$

So far we claim that the sum of a series of sinusoidal terms can approximate a periodic function. Succeeding terms have higher frequencies that are integral multiples of the fundamental. With the power series the succeeding terms had higher integral powers of the independent variable. With both series the trick is to obtain the appropriate constant coefficients. This determination with the power series depends on the function having values for all orders of derivatives at the point where the function is being expanded. To obtain the constants in the Fourier series we depend on a special property of integrals of products of sines and cosines. These integrals are summarized below. n and m are integers.

$$\int_{-\pi}^{\pi} \sin nt \, dt = 0 \qquad \int_{-\pi}^{\pi} \sin nt \cos mt \, dt = 0$$

$$\int_{-\pi}^{\pi} \cos nt \, dt = 0 \qquad \int_{-\pi}^{\pi} \sin nt \sin mt \, dt = 0 \qquad n \neq m \tag{14-14}$$

$$\int_{-\pi}^{\pi} \sin^2 nt \, dt = \pi \qquad \int_{-\pi}^{\pi} \cos nt \cos mt \, dt = 0 \qquad n \neq m$$

$$\int_{-\pi}^{\pi} \cos^2 nt \, dt = \pi$$

The sinusoidal functions are called *orthogonal* because of the property shown in the three equations with the sine-cosine products. The limits of the integrals could have been taken at the end points of any period; for instance, $\int_{-(T/2)}^{T/2}$. Even the symmetry from $-(T/2)$ to $+(T/2)$ is not necessary, although it is usually convenient.

The constant coefficients can now be evaluated by multiplying both sides of the expansion by an appropriate sine or cosine term and integrating over one period.

$$f(x) = \tfrac{1}{2}a_0 + a_1 \cos x + a_2 \cos 2x + \cdots + b_1 \sin x$$
$$+ \, b_2 \sin 2x + \cdots$$

$$\int_{-\pi}^{\pi} f(x)\, dx = \tfrac{1}{2}a_0 \int_{-\pi}^{\pi} dx + a_1 \int_{-\pi}^{\pi} \cos x\, dx + a_2 \int_{-\pi}^{\pi} \cos 2x\, dx + \cdots$$
$$+ \, b_1 \int_{-\pi}^{\pi} \sin x\, dx + b_2 \int_{-\pi}^{\pi} \sin 2x\, dx + \cdots$$
$$= \tfrac{1}{2}a_0 2\pi + 0 + 0 + \cdots + 0 + 0 + \cdots$$

Therefore,

$$a_0 = \frac{1}{\pi} \int_{-\pi}^{\pi} f(x)\, dx \qquad (14\text{--}15)$$

$$\int_{-\pi}^{\pi} f(x) \cos x\, dx = \tfrac{1}{2}a_0 \int_{-\pi}^{\pi} \cos x\, dx + a_1 \int_{-\pi}^{\pi} \cos^2 x\, dx +$$
$$a_2 \int_{-\pi}^{\pi} \cos x \cos 2x\, dx + \cdots + b_1 \int_{-\pi}^{\pi} \cos x \sin x\, dx$$
$$+ \, b_2 \int_{-\pi}^{\pi} \cos x \sin 2\, x\, dx + \cdots$$
$$= 0 + a_1 \pi + 0 + \cdots + 0 + 0 + \cdots$$

Therefore,

$$a_1 = \frac{1}{\pi} \int_{-\pi}^{\pi} f(x) \cos x\, dx \qquad (14\text{--}16)$$

Multiplying through by $\cos nx$ yields, in general,

$$a_n = \frac{1}{\pi} \int_{-\pi}^{\pi} f(x) \cos nx\, dx \qquad (14\text{--}17)$$

Multiplying through by $\sin nx$ yields, in general,

$$b_n = \frac{1}{\pi} \int_{-\pi}^{\pi} f(x) \sin nx\, dx \qquad (14\text{--}18)$$

Application to square wave. It is hard to generate a square wave, either electrically or with audio devices. The fast rise and sharp corners demand a whole range of high harmonics. Let us see how good an approximation is formed by the first few terms of a Fourier series.

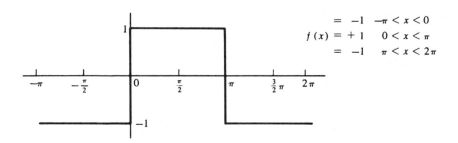

$$f(x) \begin{aligned} &= -1 &\quad -\pi < x < 0 \\ &= +1 &\quad 0 < x < \pi \\ &= -1 &\quad \pi < x < 2\pi \end{aligned}$$

The general shape of the square wave illustrated corresponds more to a sine wave than a cosine. The function is odd; i.e., $f(x) = -f(-x)$. Solving for the coefficients,

$$a_0 = \frac{1}{\pi} \int_{-\pi}^{\pi} f(x)\, dx = \frac{1}{\pi}[-1 \cdot \pi + 1 \cdot \pi] = 0$$

$$a_n = \frac{1}{\pi} \int_{-\pi}^{\pi} f(x) \cos nx\, dx = \frac{1}{\pi}\left[-\int_{-\pi}^{0} \cos nx\, dx + \int_{0}^{\pi} \cos nx\, dx \right]$$
$$= 0$$

All the a_n are zero, since $\cos nx$ is an even function, and the two terms in the brackets will always cancel.

$$b_n = \frac{1}{\pi}\left[-\int_{-\pi}^{0} \sin nx\, dx + \int_{0}^{\pi} \sin nx\, dx \right] = \frac{1}{n\pi}\left[\cos nx \Big|_{-\pi}^{0} - \cos nx \Big|_{0}^{\pi} \right]$$

$$= \frac{1}{n}\pi[2 + 2] \qquad \text{for odd } n$$

$$= \frac{1}{n}\pi[0 + 0] \qquad \text{for even } n$$

Therefore, the series for this square wave is

$$f(x) = \frac{4}{\pi} \sin x + \frac{4}{3\pi} \sin 3x + \frac{4}{5\pi} \sin 5x + \cdots \tag{14-19}$$

A plot of successive terms is shown in the diagram.

SERIES AND APPROXIMATIONS

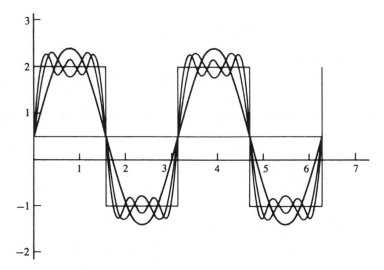

Application to sawtooth wave. The deflecting voltage on an oscilloscope or television tube has a sawtooth pattern. It can be generated by a variety of devices that produce voltages which increase linearly to some breakdown amplitude and then abruptly revert to zero and start again.

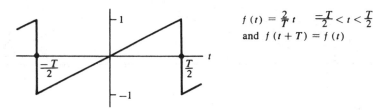

$$f(t) = \tfrac{2}{T}t \qquad \tfrac{-T}{2} < t < \tfrac{T}{2}$$
$$\text{and } f(t + T) = f(t)$$

Once again we have an odd function so that all the $a_n = 0$.

$$b_n = \frac{2}{T} \int_{-(T/2)}^{T/2} f(t) \sin \frac{2\pi}{T} nt \, dt = \frac{4}{T^2} \int_{-(T/2)}^{T/2} t \sin \frac{2\pi}{T} nt \, dt$$

Integration by parts yields

$$\int_{-(T/2)}^{T/2} t \sin \frac{2\pi}{T} nt \, dt = -\frac{T}{2\pi n} t \cos \frac{2\pi}{T} nt \Big|_{-(T/2)}^{T/2} + \frac{T}{2\pi n} \int_{-(T/2)}^{T/2} \cos \frac{2\pi}{T} nt \, dt$$

$$b_n = \frac{4}{T^2} \left[-\frac{T}{2\pi n} \frac{T}{2} \cos n\pi - \frac{T}{2\pi n} \frac{T}{2} \cos(-n\pi) + \frac{T^2}{4\pi^2 n^2} \sin \frac{2\pi}{T} nt \Big|_{-(T/2)}^{T/2} \right]$$

$$= \frac{1}{\pi n} [-\cos n\pi - \cos(-n\pi)]$$

$$= -\frac{2}{\pi n} \cos n\pi \ [\text{since } \cos(-n\pi) = +\cos(n\pi)]$$

Therefore,

$$b_n = \frac{2}{\pi n} \quad \text{if } n \text{ is odd;} \qquad b_n = -\frac{2}{\pi n} \quad \text{if } n \text{ is even}$$

The series for the sawtooth is

$$f(t) = \frac{2}{\pi} \sin \frac{2\pi}{T} t - \frac{2}{2\pi} \sin \frac{2\pi}{T} 2t + \frac{2}{3\pi} \sin \frac{2\pi}{T} 3t - \frac{2}{4\pi} \sin \frac{2\pi}{T} 4t + \cdots \quad (14\text{–}20)$$

A plot of successive terms is shown in the diagram. Note how poor the approximation is at the turnaround points.

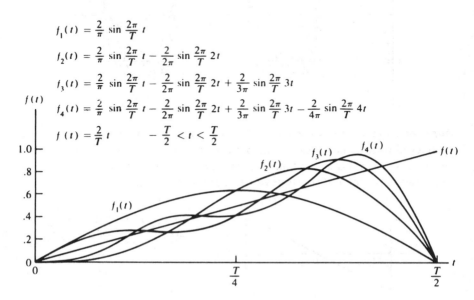

$$f_1(t) = \frac{2}{\pi} \sin \frac{2\pi}{T} t$$

$$f_2(t) = \frac{2}{\pi} \sin \frac{2\pi}{T} t - \frac{2}{2\pi} \sin \frac{2\pi}{T} 2t$$

$$f_3(t) = \frac{2}{\pi} \sin \frac{2\pi}{T} t - \frac{2}{2\pi} \sin \frac{2\pi}{T} 2t + \frac{2}{3\pi} \sin \frac{2\pi}{T} 3t$$

$$f_4(t) = \frac{2}{\pi} \sin \frac{2\pi}{T} t - \frac{2}{2\pi} \sin \frac{2\pi}{T} 2t + \frac{2}{3\pi} \sin \frac{2\pi}{T} 3t - \frac{2}{4\pi} \sin \frac{2\pi}{T} 4t$$

$$f(t) = \frac{2}{T} t \qquad -\frac{T}{2} < t < \frac{T}{2}$$

SAMPLES

1. Use the method suggested in A to find the square roots of 19, 3432 (or 34.3×10^2), 6×10^{23} (or 600×10^{21}).

2. Gravitational force in the vicinity of a spherical object (and outside its surface) is equal to $F = G(mM/r^2)$. Let $r = r_0 + h$, where h is a small distance from the surface at r_0. Expand the expression for the force in terms of h. At how many miles from the surface of the earth would your weight be less by 1 per cent?   (20 miles)

3. Carry out the expansion of $\cos x$ using Maclaurin's series and also Taylor's series around the point $x = \pi/2$.

4. Examine the curves for the power series expansion of $\sin x$ to determine the angle up to which each set of terms is good to 10 per cent; i.e., up to what angle is the first, linear, term good to 10 per cent? What angle for the first two terms?

5. Calculate the Fourier series coefficients for a triangular wave.

$$\left(f(x) = \frac{4}{\pi} \left[\sin x - \frac{1}{9} \sin 3x + \frac{1}{25} \sin 5x - \cdots \right]. \right)$$

200

CHAPTER 15 SOME COMMON
DIFFERENTIAL EQUATIONS

The problem of solving differential equations is very much like that of solving integrals. There are some common forms and techniques but no automatic method that always works. Simple differential equations can sometimes be reduced to integrals. For example,

$$\frac{dN}{dt} = -\lambda N \longrightarrow \frac{dN}{N} = -\lambda \, dt \longrightarrow \int \frac{1}{N} dN = -\lambda \int dt$$

$$\ln \frac{N}{N_0} = -\lambda t$$

$$N = N_0 e^{-\lambda t}$$

15.1 GEOMETRICAL INTERPRETATION OF DIFFERENTIAL EQUATIONS.

Frequently, it is instructive to view a differential equation as a statement about the slope and curvature of some function. The first derivative, dy/dx, is the slope of $y(x)$. The second derivative is proportional to the curvature, and for small value of slope is approximately equal to the curvature.

The *radius* of curvature (see p. 178) is

$$\rho = \frac{[1 + (dy/dx)^2]^{3 \cdot 2}}{d^2 y/dx^2}$$

The reciprocal of this is what we usually call *curvature*; a small radius implies a large curvature.

In the case of $dN/dt = -\lambda N$, we could ask, "What function is it whose first derivative, or slope, is proportional to the negative value of the function?" If we were to sketch the function following the direction given by the formula, we would draw a line with negative slope, starting out with a steep slope for large N and changing to a small slope for small N. Such a curve has the general characteristics of an exponential decay.

Consider the case of $d^2y/dt^2 = -\omega^2 y$. No direct integration is possible. Instead we ask, "What function is it whose second derivative is proportional to its negative value?" Clearly, none of the power functions has this property. Differentiating them reduces the power of their exponent. The second derivative of the exponential function is proportional to its positive value, *unless* the exponential has an imaginary argument.

$$\frac{d^2 e^{i\omega t}}{dt^2} = -\omega^2 e^{i\omega t}$$

Other functions with this property are the sine and cosine.

$$\frac{d^2 \sin \omega t}{dt^2} = -\omega^2 \sin \omega t \quad \text{or} \quad \frac{d^2 \cos \omega t}{dt^2} = -\omega^2 \cos \omega t$$

As we saw on p. 52, $e^{i\omega t}$ is also sinusoidal.

We could also have foreseen sinusoidal-type solutions by appealing to the information given in the equation about curvature of the function. This curvature is proportional to the magnitude of the function but is always positive when the function is negative, and vice versa. Such a curvature bends the function back toward equilibrium; the farther the function gets from equilibrium (the greater the absolute value), the stronger the curvature. Using this same argument, we could predict that a solution to $d^2y/dt^2 = +\omega^2 y$ must rapidly run to infinity. The curvature is in a direction to increase the function, and is proportional to that increasing function.

15.2 STEP-BY-STEP SOLUTION OF DIFFERENTIAL EQUATIONS.

Even as integrals can be solved by numerical addition of the contributions from small intervals, so differential equations can be solved step by step. For instance, the equation for free fall in a vacuum with constant acceleration is $d^2y/dt^2 = -g$. This can be solved by direct integration in two steps, or by asking "What function is it whose second derivative is a constant?" In the first case, $dy/dt = -gt + v_0$ and $y = -\frac{1}{2}gt^2 + v_0 t + y_0$. Using the second argument, the function must contain the second power of t, since differentiating that twice reduces the function to a constant.

In the real world, of course, objects fall in air, which produces a drag proportional to the square of the velocity. The differential equation for Newton's second law is

$$F_{\text{weight}} - F_{\text{drag}} = ma \longrightarrow mg - k\left(\frac{dy}{dt}\right)^2 = m\frac{d^2y}{dt^2}$$

or

$$\ddot{y} + \frac{k}{m}(\dot{y})^2 = g$$

SOME COMMON DIFFERENTIAL EQUATIONS

This equation can be solved explicitly, but let us use it to illustrate step-by-step solution. First, note that the velocity, $\dot{y}$, will increase steadily until $mg - k(\dot{y})^2 = 0$. From that point on, $\ddot{y} = 0$ and there is no further increase of velocity. The "terminal" velocity must be $\dot{y} = \sqrt{mg/k}$. A sketch of $\dot{y}$ versus time must look as sketched in the diagram. At first the object falls

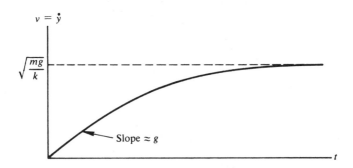

with nearly constant acceleration, g. As the velocity increases, the drag increases, and the acceleration decreases. Let us calculate the velocity and displacement during the first 10 s, using 1-s intervals. Assume that the object is a human body with a mass of 60 kg and an approximate value for k of $\frac{1}{6}$ kg/m, and let $g = 9.8 \approx 10 \ m/s^2$.

t	$\dot{y} = v_n$ $= v_{n-1}$ $+ \ddot{y}_{n-1}\Delta t$	$(\dot{y})^2$	$y_n = y_{n-1} +$ $\left(\dfrac{v_n + v_{n-1}}{2}\right)\Delta t$	v_{vacuum} $= gt$	y_{vacuum} $= \frac{1}{2}gt^2$	$\ddot{y} = g - \dfrac{k}{m}(\dot{y})^2$
0	0	0	0	0	0	$g = 10$
1	10	100	5	10	5	9.72
2	19.72	388	19.86	20	20	8.92
3	28.64	821	44.04	30	45	7.72
4	36.36	1320	76.54	40	80	6.33
5	42.69	1825	116.06	50	125	4.93
6	47.62	2265	161.21	60	180	3.71
7	51.33	2635	210.68	70	245	2.68
8	54.01	2920	263.35	80	320	1.89
9	55.90	3130	318.30	90	405	1.31
10	57.21	3275	374.85	100	500	0.90

It is assumed that during each interval the applicable acceleration is that calculated for the end of the previous interval. Each column for that interval is then filled in, in the order in which they are listed. The calculation can be done rapidly with a hand calculator. The extra significant figures were carried to

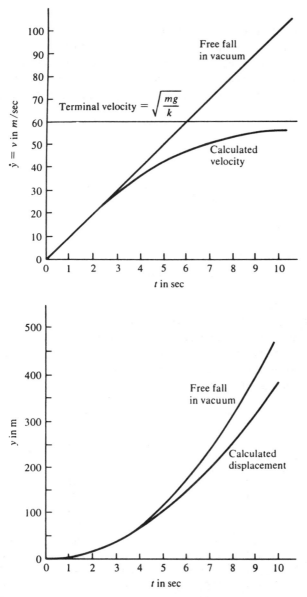

prevent cumulative error. Substitution of 10.0 for g instead of 9.8 produces only a constant 2 per cent error in both calculated and vacuum values.*

For an analytical solution of this equation, see p. 258.

15.3 DEFINITIONS CONCERNING DIFFERENTIAL EQUATIONS. An *ordinary* differential equation contains only one independent variable and total derivatives with respect to it; a *partial* differential equation contains

*For a similar treatment of a different problem, see Feynman's *Lectures on Physics*, Vol. I, Section 9–6.

more than one independent variable and partial derivatives with respect to them.

$$A\frac{d^2y}{dt^2} + B\frac{dy}{dt} + cy = 0 \qquad \text{is an ordinary differential equation}$$

$$\frac{\partial^2 V}{\partial x^2} + \frac{\partial^2 V}{\partial y^2} + \frac{\partial^2 V}{\partial z^2} = 0 \qquad \text{is a partial differential equation}$$

A *first-order* differential equation contains only first derivatives. The *order* of a differential equation is the order of the highest derivative in it. The *degree* of an equation is the number of the highest exponent of any derivative.

$$\left(\frac{d^2y}{dt^2}\right)^3 + 2\frac{dy}{dt} + y = 0$$

In this equation the order is 2 and the degree is 3. A *linear differential equation* is of first degree in the dependent variable and all its derivatives. In general a solution to a differential equation of order n has n arbitrary constants.

Consider the solutions to this second-order linear equation:

$$\frac{d^2x}{dt^2} = k^2x$$

$$x = Ae^{kt} + Be^{-kt}$$

Notice that because the differential equation is linear, if each of these terms satisfies the differential equation, their sum also satisfies it. The determination of the arbitrary constants, A and B, depends on initial or boundary conditions. At least two conditions are necessary. If at $t = 0$, $x = 0$, then $0 = A + B$ and $A = -B$. The determination of A (and B) then depends on one other condition about scale or initial slope; e.g., $x = 1$ at $t = 1$. Then $1 = A(e^k - e^{-k})$. Or if $\dot{x} = 1$ at $t = 0$, then $1 = kAe^{kt} + kAe^{kt} = 2kA$.

The final test of any solution to a differential equation is simply to plug it and its derivatives back into the equation. If the equation is satisfied, your solution is valid, although it may not be the only solution, or the general solution.

15.4 SUMMARY OF COMMON "ORDINARY" DIFFERENTIAL EQUATIONS

A. $\dfrac{dx}{dt} = K$

This equation states that the slope of a plotted function, perhaps representing a velocity, is constant. The equation can be solved by straightforward integration.

$$dx = Kdt \longrightarrow \int dx = K\int dt \longrightarrow x = Kt + C$$

At $t = 0$, $x_0 = C$, and so the general solution is $\underline{x = Kt + x_0}$.

B. $\dfrac{d^2x}{dt^2} = K$

This equation states that the second derivative, or acceleration of a plotted function is constant. This case was treated on p. 176 in terms of the acceleration of an object falling in vacuum. As also pointed out on p. 202, the solution can be guessed by seeking a function whose second derivative is constant, or by straightforward integration in two steps.

The solution is

$$x = x_0 + v_0 t + \tfrac{1}{2} K t^2$$

There are two arbitrary constants to be determined by initial conditions of x and v.

C. $\dfrac{dx}{dt} = \pm kx$ and $\dfrac{dx}{dt} = A \pm kx$

We showed examples of these equations in the section on exponential functions. They can be solved by straightforward integration, or by asking for the function whose slope is proportional to itself (or to a constant plus a quantity proportional to itself).

The solution to the first equation is

$$\frac{dx}{dt} = \pm kx \longrightarrow \frac{dx}{x} = \pm k \, dt \longrightarrow \int_{x_0}^{x} \frac{dx}{x} = \pm k \int_{0}^{t} dt \longrightarrow \ln \frac{x}{x_0}$$

$$= \pm kt \longrightarrow x = x_0 e^{\pm kt}$$

The solution to the second equation is

$$\frac{dx}{dt} = A \pm kx \longrightarrow \frac{dx}{A \pm kx} = dt \longrightarrow \int_{x_0}^{x} \frac{dx}{A \pm kx} = \int_{0}^{t} dt \longrightarrow$$

$$\pm \frac{1}{k} \ln(A \pm kx)\Big|_{x_0}^{x} = t \longrightarrow \ln\left(\frac{A \pm kx}{A \pm kx_0}\right) = \pm kt \longrightarrow A \pm kx$$

$$= (A \pm kx_0)e^{\pm kt} \longrightarrow \underline{x = \pm \frac{A}{k}(e^{\pm kt} - 1) \pm x_0 e^{\pm kt}}$$

or

$$\underline{x = \mp \frac{A}{k} \pm \left(\frac{A}{k} \pm x_0\right)e^{\pm kt}}$$

Compare these solution forms to those obtained by fabricating a solution from plausibility arguments as done on p. 58. To translate the symbols, $x = p$, $x_0 = P$, $A = -K$, $\pm k = +r$, and when $x = p \to 0$, $t \to T$. Then

$$p = \frac{K}{r} + \left(P - \frac{K}{r}\right)e^{rt}$$

As shown on p. 59,

$$K = rP \frac{1}{1 - e^{-rT}}$$

Substituting,

$$p = \frac{P}{1 - e^{-rT}} + \left(P - \frac{P}{1 - e^{-rT}}\right)e^{rt}$$

$$p = P\frac{1 - e^{-rT}e^{rt}}{1 - e^{-rT}} = P\frac{e^{rT} - e^{rt}}{e^{rT} - 1}$$

which is the expression on p. 57.

D. $\dfrac{d^2x}{dt^2} = -kx$

This is the equation yielding simple harmonic motion. The acceleration is proportional to the displacement and in the opposite direction, which provides a restoring effect. We have discussed solutions to the equation on pp. 46–49.

Equivalent forms of the general solution are

$$x = A \sin \sqrt{k}\, t + B \cos \sqrt{k}\, t$$
$$x = A \sin \sqrt{k}\, (t + \alpha) \quad \text{or} \quad A \cos \sqrt{k}\, (t + \beta)$$
$$x = A e^{i\sqrt{k}\, t} + B e^{-i\sqrt{k}\, t}$$
$$x = A e^{\pm i\sqrt{k}\, (t + \alpha)}$$

The arbitrary constants A and B are the amplitudes of the sinusoidal oscillations, determined by the initial conditions. Alternatively, the conditions may specify an amplitude A and a phase constant, α, which is determined by the definition of t_0.

E. $\dfrac{d^2x}{dt^2} = +kx$

The acceleration, or curvature, in this case is in the same direction as the displacement. The magnitude of the variable rapidly gets out of hand.

The general solution is

$$x = A e^{\sqrt{k}\, t} + B e^{-\sqrt{k}\, t}$$

or

$$x = A e^{\pm \sqrt{k}\, (t + \alpha)}$$

Notice that the solution is exponential, as is the solution to $dx/dt = +kx$. For the exponential, both the slope and the curvature are proportional to the function.

F. $\dfrac{d^2x}{dt^2} = -k \sin x$

On p. 48 we analyzed the dynamics of a simple pendulum and obtained

$d^2\theta/dt^2 = -\omega^2 \sin \theta$, where $\omega = \sqrt{g/L}$. We then made the approximation that for small θ, $\sin \theta \approx \theta$. This assumption reduces the differential equation to the much simpler one for simple harmonic motion.

There is no explicit solution to the equation when the second derivative is proportional to the negative of the sine of the function. The solution involves elliptic integrals that must be solved numerically.

Here are some steps that reduce the original equation to an elliptic integral:

Multiply both sides by $2(dx/dt)$:

$$2\frac{dx}{dt}\frac{d^2x}{dt^2} = -2k\frac{dx}{dt}\sin x$$

Integrate to obtain

$$\left(\frac{dx}{dt}\right)^2 = 2k \cos x + C$$

If $dx/dt = 0$ when $x = x_0$ (velocity is zero at maximum amplitude),

$$0 = 2k \cos x_0 + C$$

Therefore,

$$\left(\frac{dx}{dt}\right)^2 = 2k(\cos x - \cos x_0)$$

The integral of this is

$$t = \sqrt{\frac{1}{2k}} \int \frac{dx}{\sqrt{\cos x - \cos x_0}}$$

Carrying out a series approximation to this integral yields the period

$$T = 2\pi\sqrt{\frac{1}{k}}\left(1 + \frac{1}{4}\sin^2 \frac{x_0}{2} + \frac{9}{64}\sin^4 \frac{x_0}{2} + \cdots\right)$$

Notice that the period depends on the amplitude. If $x_0 = 30°$, the first term correction is $\frac{1}{4}\sin^2 15° = 0.017$, or 1.7 per cent of the main term. The period of a simple pendulum is independent of amplitude within 1 per cent for angles up to 23°.

G. $\dfrac{d^2x}{dt^2} + k\dfrac{dx}{dt} = K$

This is the equation for acceleration equal to a constant, minus a friction term proportional to the velocity. Such a linear dependence on velocity is characteristic of the drag of objects falling slowly through a viscous fluid. Newton's second law for such a case would be

$$F_{down} - F_{friction} = ma \text{ and (weight − buoyancy)} - kv = m\ddot{x}$$

This is equivalent to the form of the equation above. To emphasize the physical significance, we shall analyze

$$m\frac{d^2x}{dt^2} + k\frac{dx}{dt} = W$$

Reduce the order of the equation by letting $dx/dt = v$:

$$m\frac{dv}{dt} + kv = W$$

Separate the variables and integrate:

$$m\,dv + kv\,dt = W\,dt \longrightarrow dv + \frac{k}{m}v\,dt = \frac{W}{m}\,dt \longrightarrow$$

$$dv = (W - kv)\frac{1}{m}\,dt = \left(\frac{W}{k} - v\right)\frac{k}{m}\,dt$$

$$\frac{dv}{v - (W/k)} = -\frac{k}{m}\,dt \longrightarrow \int_{v_0}^{v}\frac{dv}{v - (W/k)} = -\frac{k}{m}\int_{0}^{t}dt$$

$$\ln\left(\frac{v - (W/k)}{v_0 - (W/k)}\right) = -\frac{k}{m}t \longrightarrow \ln\left[\frac{(W/k) - v}{W/k}\right] = -\frac{k}{m}t \qquad \text{if } v_0 = 0$$

$$\text{when } t = 0.$$

$$\frac{W}{k} - v = \frac{W}{k}e^{-(k/m)t} \longrightarrow v = \frac{dx}{dt} = \frac{W}{k}[1 - e^{-(k/m)t}]$$

$$x = \frac{W}{k}t + \frac{Wm}{k^2}e^{-(k/m)t} + x_0$$

If $x = 0$ when $t = 0$, $x_0 = -Wm/k^2$:

$$x = \frac{W}{k}t - \frac{Wm}{k^2}[1 - e^{-(k/m)t}]$$

Note that at $t = 0$, $x = 0$, and as $t \to \infty$, $x \to (W/k)t - (Wm/k^2)$. The velocity for large t is essentially constant and equal to W/k.

H. $A\dfrac{d^2x}{dt^2} + B\dfrac{dx}{dt} + Cx = 0$

This is a linear differential equation with constant coefficients. Sometimes it is written in operator notation where $D = d/dt$ and $D^2 = d^2/dt^2$:

$$AD^2x + BDx + Cx = 0$$

Such an equation is always satisfied by an exponential solution:

$$x = c_1 e^{\alpha t}$$

Substituting in the equation,

$$A\alpha^2 c_1 e^{\alpha t} + B\alpha c_1 e^{\alpha t} + Cc_1 e^{\alpha t} = 0$$

$$\therefore A\alpha^2 + B\alpha + C = 0$$

The differential equation is satisfied if

$$\alpha = \frac{-B \pm \sqrt{B^2 - 4AC}}{2A}$$

To give physical significance to this solution, change the symbols of the equation so that it describes a closed electric circuit with inductance, L, capacitance, C, and resistance, R. The sum of the voltage differences around the circuit must be zero.

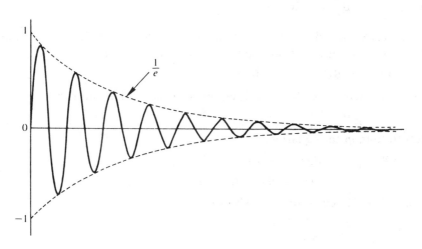

$$\Delta V = -L\frac{di}{dt} = L\frac{d^2q}{dt^2}$$

$$\Delta V = \frac{q}{C}$$

$$\Delta V = -iR = R\frac{dq}{dt}$$

$$L\frac{d^2q}{dt^2} + R\frac{dq}{dt} + \frac{q}{C} = 0$$

The general solution to this equation is

$$q = c_1 e^{(-R+\sqrt{R^2-4(L/C)}/2L)t} + c_2 e^{(-R-\sqrt{R^2-4(L/C)}/2L)t}$$

or

$$q = c_1 e^{-(R/2L)t} e^{\sqrt{(R^2/4L^2)-(1/LC)}\,t} + c_2 e^{-(R/2L)t} e^{-\sqrt{(R^2/4L^2)-(1/LC)}\,t}$$

Each term contains a damping factor, $e^{-(R/2L)t}$, and another exponential that may be real, or imaginary and thus sinusoidal. If $(1/LC) < (R^2/4L^2)$, the charge diminishes to zero exponentially, the stored energy being dissipated in the resistance. If $(1/LC) > (R^2/4L^2)$, the charge flows back and forth around the circuit, first storing energy in the capacitance and then in the inductance.

SOME COMMON DIFFERENTIAL EQUATIONS

A graph of damped oscillation is shown in the diagram. The time constant for decay to $1/e$ of the starting amplitude is $\tau = 2L/R$. If the resistance were zero, the solution would be

$$q = c_1 e^{i(\sqrt{1/LC}\, t + \alpha)} \qquad \text{or} \qquad c_1 \sin\left(\sqrt{\frac{1}{LC}}t + \alpha\right)$$

The angular frequency would be $\omega_0 = \sqrt{1/LC}$.

The frequency with resistance can be found by examining the argument of the exponential, remembering that we assume $(1/LC) > (R^2/4L^2)$:

$$e^{\sqrt{(R^2/4L^2)-(1/LC)}\, t} = e^{i\sqrt{(1/LC)-(R^2/4L^2)}\, t} = e^{i\omega t}$$

$$\omega = \sqrt{\omega_0^2 - \frac{R^2}{4L^2}} = \sqrt{\frac{4\pi^2}{T_0^2} - \frac{1}{\tau^2}}$$

The frequency is reduced by the presence of the resistance. In terms of the sine function, a complete solution for the oscillatory mode is

$$q = c_1 e^{-(R/2L)t} \sin\left(\sqrt{\frac{1}{LC} - \frac{R^2}{4L^2}}t + \alpha\right)$$

All these equations apply equally well to a mechanical oscillating system. In that case, each of the terms in the original differential equation has the dimensions of a force:

$$m\frac{d^2x}{dt} + c\frac{dx}{dt} + kx = 0$$

The first term is mass times acceleration. The second is a friction term, where the friction is proportional to the velocity, dx/dt. The proportionality constant might be a function of the shape of the object. The third term is the restoring force, proportional to the displacement from equilibrium, x, with a "spring" constant, k.

I. $A\dfrac{d^2x}{dt} + B\dfrac{dx}{dt} + Cx = f(t)$

This is the equation for a forced oscillator. The system is not free to oscillate at its normal frequency, or to decay at its own rate, but must follow the imposed function, $f(t)$.

To show an example that is familiar and physically meaningful, let us use the symbols for a series circuit with inductance, capacitance, and resistance, and impose a sinusoidal driving voltage. The equation becomes

$$L\frac{d^2q}{dt^2} + R\frac{dq}{dt} + \frac{q}{C} = Ee^{i\omega t}$$

The exponential form of the sinusoidal function is easier to manipulate. We can extract the real part of the solution in terms of a sine function at the end of the calculation.

If we can find a particular solution that satisfies the complete equation, we can then add to it the general solution for the free equation—since that solution makes the left side zero. For a particular solution, notice that the final function of q must oscillate with the same frequency as the driving function, though not necessarily in phase. Let $q(t) = Qe^{i\omega t}$. Substituting into the equation,

$$-LQ\omega^2 e^{i\omega t} + iRQ\omega e^{i\omega t} + \frac{1}{C}Qe^{i\omega t} = Ee^{i\omega t}$$

The term, $e^{i\omega t}$, cancels out. We are left with real and imaginary terms. Solve for the complex amplitude, Q.

$$Q = \frac{E}{(1/C) - L\omega^2 + i\omega R} = \frac{E/L}{(1/LC) - \omega^2 + i\omega(R/L)}$$

$$= \frac{E/L}{(\omega_0^2 - \omega^2) + i\omega(R/L)}$$

The frequency, ω_0, is the normal frequency of the undamped, undriven system. $\omega_0 = \sqrt{1/LC}$. Q can be expressed in real and imaginary parts by multiplying numerator and denominator by the complex conjugate of the denominator.

$$Q = \frac{(E/L)[(\omega_0^2 - \omega^2) - i\omega(R/L)]}{[(\omega_0^2 - \omega^2) + i\omega(R/L)][(\omega_0^2 - \omega^2) - i\omega(R/L)]}$$

$$= \frac{(E/L)(\omega_0^2 - \omega^2)}{(\omega_0^2 - \omega^2)^2 + \omega^2(R^2/L^2)} - i\frac{(E/L)(\omega R/L)}{(\omega_0^2 - \omega^2)^2 + \omega^2(R^2/L^2)}$$

Picture this complex amplitude in terms of a vector in the complex plane.

$$\frac{\dfrac{E}{L}(\omega_0^2 - \omega^2)}{(\omega_0^2 - \omega^2)^2 + \omega^2 \dfrac{R^2}{L^2}}$$

$$\tan\phi = \frac{-\omega\dfrac{R}{L}}{(\omega_0^2 - \omega^2)}$$

Imaginary

Real

$$\frac{\dfrac{E}{L}\dfrac{\omega R}{L}}{(\omega_0^2 - \omega^2)^2 + \omega^2 \dfrac{R^2}{L^2}}$$

$$\frac{\dfrac{E}{L}}{\sqrt{(\omega_0^2 - \omega^2)^2 + \omega^2 \dfrac{R^2}{L^2}}}$$

$$Q = \frac{\dfrac{E}{L}(\cos\phi - i\sin\phi)}{\sqrt{(\omega_0^2 - \omega^2)^2 + \omega^2 \dfrac{R^2}{L^2}}} = \frac{\dfrac{E}{L}e^{-i\phi}}{\sqrt{(\omega_0^2 - \omega^2)^2 + \omega^2 \dfrac{R^2}{L^2}}}$$

The particular solution is

$$q = \frac{(E/L)e^{i(\omega t - \phi)}}{\sqrt{(\omega_0^2 - \omega^2)^2 + \omega^2(R^2/L^2)}}$$

The complete solution is the sum of the general and particular solutions.

$$q = Ce^{-(R/2L)t}\sin\left(\sqrt{\frac{1}{LC} - \frac{R^2}{4L^2}}\,t + \alpha\right) + \frac{E/L}{\sqrt{(\omega_0^2 - \omega^2)^2 + \omega^2(R^2/L^2)}}e^{i(\omega t - \phi)}$$

The complete solution satisfies the equation because when substituted in, the general solution yields zero and the particular solution yields the driving term. The general solution is a transient that eventually damps out with a time constance $\tau = 2L/R$. After that the particular solution dominates. Note first that although the system oscillates at the driving frequency ω, it is not in phase with it. The phase angle ϕ and the amplitude depend on the relationship of the driving frequency ω and the natural system frequency, ω_0. When $\omega = \omega_0$, $\tan\phi = \infty$ and $\phi = 90°$. For ω smaller than ω_0, $\phi > 90°$. For ω greater than ω_0, $\phi < 90°$. If the driving frequency is close to the natural frequency, the amplitude rises. The maximum amplitude occurs when the denominator is a minimum. If ω_0 is varied, the resonance is reached when $\omega_0 = \omega$. Usually, the imposed frequency is swept to obtain resonance. In that case, both terms of the denominator are varying. To find the resonance point, differentiate the quantity inside the square root by ω.

$$-4\omega(\omega_0^2 - \omega^2) + 2\omega\frac{R^2}{L^2} = 0 \qquad 2(\omega_0^2 - \omega^2) = \frac{R^2}{L^2}$$

$$\omega_{\text{resonance}} = \sqrt{\omega_0^2 - \frac{R^2}{2L^2}}$$

If the damping constant, $R/2L$, is small compared with ω_0, the resonance frequency is approximately equal to ω_0. The diagram shows the dependence of amplitude on the ratio of ω/ω_0 for three different values of $R/2L$.

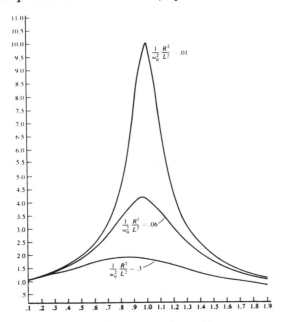

J. $\dfrac{d^2x_1}{dt^2} + \dfrac{K}{m}x_1 + \dfrac{k}{m}(x_1 - x_2) = 0$ (1)

$\dfrac{d^2x_2}{dt^2} + \dfrac{K}{m}x_2 + \dfrac{k}{m}(x_2 - x_1) = 0$ (2)

These are the equations of two identical oscillators coupled together. Each has a mass, m, and a spring constant, K. They are coupled loosely with some mechanism that acts like a spring with constant, k. This approximately describes a system such as two pendulums supported side by side from a slack horizontal rope, or lightly fastened together with a spring from bob to bob. The coupling effect depends on the displacement $(x_1 - x_2)$ between the two objects. Note that if the objects vibrate together in such a way that $x_1 - x_2$ is always zero, we are left simply with identical equations describing identical simple harmonic motion for the two objects.

The general solution proceeds by rearranging terms and assuming sinusoidal solutions with undetermined amplitudes and frequency. As usual with lengthy derivations, it is easier to use the exponential sinusoidal, $e^{i\omega t}$, instead of $\sin \omega t$.

$\dfrac{d^2x_1}{dt^2} + \left(\dfrac{K+k}{m}\right)x_1 - \dfrac{k}{m}x_2 = 0$ (1) rearranged

$\dfrac{d^2x_2}{dt^2} + \left(\dfrac{K+k}{m}\right)x_2 - \dfrac{k}{m}x_1 = 0$ (2) rearranged

Let $x_1 = A_1 e^{i\omega t}$ and $x_2 = A_2 e^{i\omega t}$.

$-\omega^2 A_1 e^{i\omega t} + \left(\dfrac{K+k}{m}\right)A_1 e^{i\omega t} - \dfrac{k}{m}A_2 e^{i\omega t} = 0$ (1) solution substituted

$-\omega^2 A_2 e^{i\omega t} + \left(\dfrac{K+k}{m}\right)A_2 e^{i\omega t} - \dfrac{k}{m}A_1 e^{i\omega t} = 0$ (2) solution substituted

$\left(-\omega^2 + \dfrac{K+k}{m}\right)A_1 - \dfrac{k}{m}A_2 = 0$ (1) solution rearranged

$-\dfrac{k}{m}A_1 + \left(-\omega^2 + \dfrac{K+k}{m}\right)A_2 = 0$ (2) solution rearranged

This pair of simultaneous equations has nonzero solutions for A_1 and A_2 only if

$\left(-\omega^2 + \dfrac{K+k}{m}\right)^2 - \left(\dfrac{k}{m}\right)^2 = 0$

[This is true because under such conditions the determinant must equal zero. Alternatively, solve by multiplying the numerator by k/m and the denominator by $(-\omega^2 + (K+k)/m)$.]

$\therefore \quad -\omega^2 + \dfrac{K+k}{m} = \pm\dfrac{k}{m}$

$\omega^2 = \dfrac{K+k\pm k}{m} = \dfrac{K+2k}{m}$ or $\dfrac{K}{m}$

SOME COMMON DIFFERENTIAL EQUATIONS

Apparently, two frequencies are possible, $\omega_a = \sqrt{(K + 2k)/m}$ and $\omega_b = \sqrt{K/m}$. The slower frequency, ω_b, is the same frequency that the independent systems would have if there were no coupling.

Substituting these two solutions for ω into the equations for A_1 and A_2, we get

For ω_b: $\left(-\dfrac{K}{m} + \dfrac{K + k}{m} \right) A_1 - \dfrac{k}{m} A_2 = 0 \longrightarrow A_1 = A_2$

For ω_a: $\left(-\dfrac{K + 2k}{m} + \dfrac{K + k}{m} \right) A_1 - \dfrac{k}{m} A_2 = 0 \longrightarrow A_1 = -A_2$

One solution is

$$x_1 = A e^{i\omega_b t} \qquad x_2 = A e^{i\omega_b t}$$

and the other is

$$x_1 = A e^{i\omega_a t} \qquad x_2 = -A e^{i\omega_a t}$$

In the first case, b, the objects vibrate in phase at the uncoupled frequency. In the second case, a, the objects are 180° out of phase and vibrate at a higher frequency. Although the system *can* oscillate in either of these two modes,

ω_b

$A_1 = A_2$

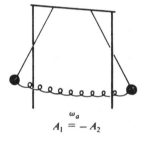

ω_a

$A_1 = -A_2$

usually there is a mixture that appears quite different. To see how this mixture arises, let us use two other variables besides x_1 and x_2. We introduce their sum and their difference:

$$\lambda_a = x_1 - x_2 \qquad x_1 = \dfrac{\lambda_a + \lambda_b}{2}$$

$$\lambda_b = x_1 + x_2 \qquad x_2 = \dfrac{\lambda_b - \lambda_a}{2}$$

Substitute these into the original differential equations and solve.

$$\dfrac{1}{2}\dfrac{d^2\lambda_a}{dt^2} + \dfrac{1}{2}\dfrac{d^2\lambda_b}{dt^2} + \left(\dfrac{K + k}{m}\right)\left(\dfrac{\lambda_a + \lambda_b}{2}\right) - \dfrac{k}{m}\left(\dfrac{\lambda_b - \lambda_a}{2}\right) = 0$$

$$\dfrac{1}{2}\dfrac{d^2\lambda_b}{dt^2} - \dfrac{1}{2}\dfrac{d^2\lambda_a}{dt^2} + \left(\dfrac{K + k}{m}\right)\left(\dfrac{\lambda_b - \lambda_a}{2}\right) - \dfrac{k}{m}\left(\dfrac{\lambda_a + \lambda_b}{2}\right) = 0$$

Canceling out terms, these yield

$$\frac{d^2\lambda_b}{dt^2} + \frac{K}{m}\lambda_b = 0$$

$$\frac{d^2\lambda_a}{dt^2} + \frac{K + 2k}{m}\lambda_a = 0$$

These are standard equations for simple harmonic motion. The variables, λ_a and λ_b, have "uncoupled" the equations. They are called "normal" coordinates, and the two frequencies are characteristic of the "normal" modes.

$$\lambda_b = A_b e^{i\omega_b t} \qquad \lambda_a = A_a e^{i\omega_a t}$$

$$\omega_b = \sqrt{\frac{K}{m}} \qquad \omega_a = \sqrt{\frac{K + 2k}{m}}$$

In terms of these constants, we can retrieve x_1 and x_2:

$$x_1 = \tfrac{1}{2}A_a e^{i\omega_a t} + \tfrac{1}{2}A_b e^{i\omega_b t}$$

$$x_2 = \tfrac{1}{2}A_b e^{i\omega_b t} - \tfrac{1}{2}A_a e^{i\omega_a t}$$

Now let $A_a = A_1 - A_2$ and $A_b = A_1 + A_2$. Then

$$x_1 = \tfrac{1}{2}[A_1 e^{i\omega_a t} - A_2 e^{i\omega_a t} + A_1 e^{i\omega_b t} + A_2 e^{i\omega_b t}]$$

$$x_2 = \tfrac{1}{2}[A_1 e^{i\omega_b t} + A_2 e^{i\omega_b t} - A_1 e^{i\omega_a t} + A_2 e^{i\omega_a t}]$$

Next define an average frequency, $\bar{\omega}$, and a difference of frequency, $\Delta\omega$.

$$\bar{\omega} = \frac{\omega_a + \omega_b}{2} \qquad \Delta\omega = \omega_a - \omega_b$$

In terms of these frequencies, the expressions, for x_1 and x_2 become

$$x_1 = \tfrac{1}{2}A_1 e^{i\bar{\omega}t}[e^{i(\Delta\omega/2)t} + e^{-i(\Delta\omega/2)t}] - \tfrac{1}{2}A_2 e^{i\bar{\omega}t}[e^{i(\Delta\omega/2)t} - e^{-i(\Delta\omega/2)t}]$$

$$x_2 = \tfrac{1}{2}A_1 e^{i\bar{\omega}t}[-e^{i(\Delta\omega/2)t} + e^{-i(\Delta\omega/2)t}] + \tfrac{1}{2}A_2 e^{i\bar{\omega}t}[+e^{i(\Delta\omega/2)t} + e^{-i(\Delta\omega/2)t}]$$

Since $(e^{i\theta} + e^{-i\theta})/2 = \cos\theta$ and $(e^{i\theta} - e^{-i\theta})/2i = \sin\theta$,

$$x_1 = A_1 e^{i\bar{\omega}t}\left(\cos\frac{\Delta\omega}{2}t\right) - iA_2 e^{i\bar{\omega}t}\left(\sin\frac{\Delta\omega}{2}t\right)$$

$$x_2 = -iA_1 e^{i\bar{\omega}t}\left(\sin\frac{\Delta\omega}{2}t\right) + A_2 e^{i\bar{\omega}t}\left(\cos\frac{\Delta\omega}{2}t\right)$$

We can simplify these expressions by assuming some initial conditions and thus determining the constants. If $x_1(0) = x_0$ and $x_2(0) = 0$, and if the initial velocities are zero (you obtain these conditions by pulling back one of the pendulums and letting it go), $A_1 = x_0$ and $A_2 = 0$. Under these conditions

$$x_1 = x_0 e^{i\bar{\omega}t} \cos\frac{\Delta\omega}{2}t$$

$$x_2 = -ix_0 e^{i\bar{\omega}t} \sin\frac{\Delta\omega}{2}t$$

Remember that $e^{i\theta} = \cos\theta + i\sin\theta$. Substitute and take only the real parts.

$$x_1 = \left(x_0 \cos\frac{\Delta\omega}{2}t\right)\cos\bar{\omega}t$$

$$x_2 = \left(x_0 \sin\frac{\Delta\omega}{2}t\right)\sin\bar{\omega}t$$

Each object oscillates with a frequency $\bar{\omega}$, which is an average of the two normal mode frequencies. These oscillations are 90° out of phase (one is a sine term and the other a cosine term), and are modulated by time-dependent amplitudes that are sinusoidal with a frequency of

$$\frac{1}{2}\Delta\omega = \frac{\omega_a - \omega_b}{2} = \frac{1}{2}\left(\sqrt{\frac{K+2k}{m}} - \sqrt{\frac{K}{m}}\right)$$

If $k \ll K$, this frequency of the modulating amplitude is small compared with the main frequency. The pendulum that was originally disturbed gradually swings in smaller arcs, transferring its energy to the second one, and eventually it stops momentarily. Meanwhile the amplitude of the second one steadily increases until it is at a maximum when the first one has stopped. Then the process repeats, with the energy being fed back into the first one. Increasing the strength of the coupling increases the frequency of transfer of energy from number one to number two and back again.

K. For the equation $\dfrac{d^2x}{dt^2} + k\left(\dfrac{dx}{dt}\right)^2 = K$

See Appendix 12 on p. 258.

CHAPTER16 DIFFERENTIAL OPERATORS

Things change. We describe these changes mathematically in terms of de-rivatives. A change of a variable, Φ, with respect to time is simply $d\Phi/dt$. If the variable is also a function of position, the change in the variable depends on the direction. $\partial\Phi/\partial x$ may be different from $\partial\Phi/\partial y$.

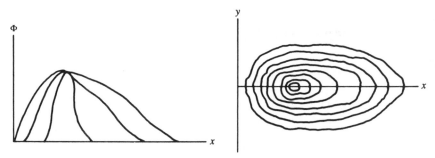

16.1 THE GRADIENT., The perspective drawing shows a hill modeling a scalar function $\Phi(x, y)$. This function might represent gravitational or electric potential or the temperature. The contour map is a plot of lines of constant Φ. At any given point the slope depends on the direction.

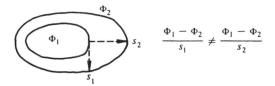

$$\frac{\Phi_1 - \Phi_2}{s_1} \neq \frac{\Phi_1 - \Phi_2}{s_2}$$

The maximum slope at any particular point is given by $\partial\Phi/\partial n$, where n is the normal to the curve of constant Φ. In the case of electric potential, $V(x, y)$, this slope defines the electric field: $\mathbf{E} = -(\partial V/\partial n)$. The electric field lines are perpendicular to the equipotential lines.

The change of the function in three dimensions with respect to an arbitrary direction, s, is

$$\frac{d\Phi}{ds} = \frac{\partial\Phi}{\partial x}\frac{dx}{ds} + \frac{\partial\Phi}{\partial y}\frac{dy}{ds} + \frac{\partial\Phi}{\partial z}\frac{dz}{ds}$$

$$= \frac{\partial\Phi}{\partial x}\cos(s, x) + \frac{\partial\Phi}{\partial y}\cos(s, y) + \frac{\partial\Phi}{\partial z}\cos(s, z) \qquad (16\text{--}1)$$

This expression is the same as that for the component of a vector in the direction s. For instance, if $\mathbf{F} = \hat{i}F_x + \hat{j}F_y + \hat{k}F_z$, the component of $\mathbf{F}$ in the s direction is

$$F_s = F_x\cos(s, x) + F_y\cos(s, y) + F_z\cos(s, z)$$

It appears that $d\Phi/ds$ must be the component in the s direction of a vector equal to $\hat{i}(\partial\Phi/\partial x) + \hat{j}(\partial\Phi/\partial y) + \hat{k}(\partial\Phi/\partial z)$. This vector must be equal to $d\Phi/dn$, since the maximum value of a component of a vector must be the vector itself, and the maximum slope must be in the direction of the perpendicular to the contour line of constant Φ.

$$\frac{d\Phi}{dn} = \hat{i}\frac{\partial\Phi}{\partial x} + \hat{j}\frac{\partial\Phi}{\partial y} + \hat{k}\frac{\partial\Phi}{\partial z} \qquad (16\text{--}2)$$

This vector of maximum slope is called the gradient and is symbolized either by $grad$ or ∇:

$$\text{grad } \Phi = \nabla\Phi = \frac{d\Phi}{dn} = \hat{i}\frac{\partial\Phi}{\partial x} + \hat{j}\frac{\partial\Phi}{\partial y} + \hat{k}\frac{\partial\Phi}{\partial z} \qquad (16\text{--}3)$$

The direction of the gradient is the direction of the steepest ascent in Φ, and the magnitude is the magnitude of the slope in that direction.

It is convenient to consider ∇ (pronounced "del") by itself as a differential operator:

$$\nabla = \hat{i}\frac{\partial}{\partial x} + \hat{j}\frac{\partial}{\partial y} + \hat{k}\frac{\partial}{\partial z} \qquad (16\text{--}4)$$

It is not a vector by itself, but is an operator on a function, in this case a scalar. The operation produces a vector from a scalar function.

16.2 THE DIVERGENCE. The ∇ operator can also be defined to operate on a vector. Consider first the significance of an operation in the form of a dot (or scalar) product:

$$\nabla\cdot\mathbf{F} = \left(\hat{i}\frac{\partial}{\partial x} + \hat{j}\frac{\partial}{\partial y} + \hat{k}\frac{\partial}{\partial z}\right)\cdot(\hat{i}F_x + \hat{j}F_y + \hat{k}F_z) = \frac{\partial F_x}{\partial x} + \frac{\partial F_y}{\partial y} + \frac{\partial F_z}{\partial z}$$

Note that $\nabla\Phi$ forms a vector, but $\nabla\cdot\mathbf{F}$ yields a scalar. The operator $\nabla\cdot\mathbf{F}$ is called the divergence of F or div $\mathbf{F}$.

The physical significance of the divergence is more evident from its alternative definition.

$$\text{div } \mathbf{F} = \lim_{\Delta\tau\to 0} \frac{\int_{\Delta s} F_n \, ds}{\Delta\tau} = \lim_{\Delta\tau\to 0} \frac{\int_s \mathbf{F}\cdot d\mathbf{s}}{\Delta\tau} \tag{16-6}$$

The numerator is a surface integral of the normal vector taken over a closed region with volume $\Delta\tau$. For every element of surface area of the closed

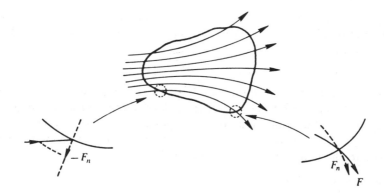

volume, the component of the vector field that is normal to the surface is multiplied by the surface area. The product is sometimes positive and sometimes negative, depending on whether the vector lines are leaving the volume or entering it. The divergence of the vector field is defined as the limit of the integral of all these products taken over the whole volume, and divided by the volume as the volume goes to zero.

The lines representing the loci of vectors in the field are called *flux lines*. The flux through a surface is defined as

$$\Phi_F = \int_s F\cdot ds \tag{16-7}$$

The strength of the field at a point is thus given as

$$F = \frac{\Delta\Phi_F}{\Delta s} \tag{16-8}$$

or, in words, the number of flux lines per unit area. The divergence is a measure of the *net* flux entering and leaving a region. If the divergence is zero, there is continuity of these lines. All the lines that went in, came out. If the divergence is not zero, there must be a source or sink for the flux. If the flux is composed of electric lines of force, then div $\mathbf{E} = \nabla\cdot\mathbf{E} = 0$ in a region where there are no electric charges. At a point where there is a charge, $\nabla\cdot\mathbf{E} = K\sigma_E$. For gravitation, $\nabla\cdot\mathbf{g} = K\sigma_m$, where σ_m is the gravitational

DIFFERENTIAL OPERATORS

"charge density." For the magnetic field, $\nabla \cdot \mathbf{B} = 0$, since there are no isolated magnetic charges (so far as we know).

The relationship between the integral and differential definitions of divergence can be seen by constructing a rectangular volume element in a hypothetical vector field. The approximate amount of flux entering the left-

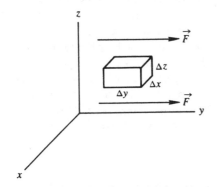

hand face is $-F_y \Delta x \Delta z$. Leaving the right-hand face is $+F_{y+\Delta y} \Delta x \Delta z$. The net flow through the volume in the xz plane is $(F_{y+\Delta y} - F_y) \Delta x \Delta z \longrightarrow$ $(\partial F_y / \partial y)\, dy\, dx\, dz$ in the limit as the volume element $\longrightarrow 0$. Similarly, the net flow through the top and side of the volume is

$$\frac{\partial F_z}{\partial z}\, dz\, dx\, dy \qquad \text{and} \qquad \frac{\partial F_x}{\partial x}\, dx\, dy\, dz$$

Since $\lim \Delta\tau = dx\, dy\, dz$,

$$\text{div } \mathbf{F} = \lim_{\Delta\tau \to 0} \frac{\int_s \mathbf{F} \cdot d\mathbf{s}}{\Delta\tau} = \frac{\partial F_x}{\partial x} + \frac{\partial F_y}{\partial y} + \frac{\partial F_z}{\partial z} = \nabla \cdot \mathbf{F} \qquad (16\text{--}9)$$

16.3 THE CURL. The del operator can also perform a vector product operation on a vector. This is called the *curl*.

$$\text{curl } \mathbf{F} = \nabla \times \mathbf{F} = \left(\hat{i}\frac{\partial}{\partial x} + \hat{j}\frac{\partial}{\partial y} + \hat{k}\frac{\partial}{\partial z} \right) \times (\hat{i}F_x + \hat{j}F_y + \hat{k}F_z)$$

$$= \begin{vmatrix} \hat{i} & \hat{j} & \hat{k} \\ \dfrac{\partial}{\partial x} & \dfrac{\partial}{\partial y} & \dfrac{\partial}{\partial z} \\ F_x & F_y & F_z \end{vmatrix} = \left(\frac{\partial F_z}{\partial y} - \frac{\partial F_y}{\partial z} \right)\hat{i} + \left(\frac{\partial F_x}{\partial z} - \frac{\partial F_z}{\partial x} \right)\hat{j}$$

$$+ \left(\frac{\partial F_y}{\partial x} - \frac{\partial F_x}{\partial y} \right)\hat{k} \qquad (16\text{--}10)$$

The curl is itself a vector, perpendicular to the vector $\mathbf{F}$. The name stems from hydrodynamics, referring to a region of liquid that is rotating around an internal axis that may itself be moving. The significance of this name is evident from the definition of the curl in terms of an integral:

$$\text{curl } \mathbf{F} = \lim_{\Delta s \to 0} \frac{\int_c F_l \, dl}{\Delta s} \qquad (16\text{-}11)$$

The numerator contains a line or contour integral around a closed loop in a vector field. The component of the vector F in the direction of the loop is multiplied by the line element and summed around the loop. The result is "normalized" by dividing by the area, Δs, enclosed by the loop. The curl at a point is the limit of this ratio as the loop shrinks to the point. When the plane of the loop is chosen to give the largest value of the integral, the direction of the curl is perpendicular to the plane of the loop. If curl $\mathbf{F} = 0$ for every point in the field, then $\mathbf{F}$ is defined to be *irrotational*. This is necessarily the case if $\mathbf{F}$ is derived from a potential, $\mathbf{F} = \nabla \Phi$. This type of field is called "conservative."

$$\text{curl (grad } \Phi) \equiv \nabla \times \nabla \Phi \equiv 0 \qquad (16\text{-}12)$$

In this case a line integral between any two points is independent of the path. A closed loop line integral is therefore zero.

The relationship between the integral and differential definitions of curl can be seen by constructing a rectangular loop in a vector field and performing the contour integral.

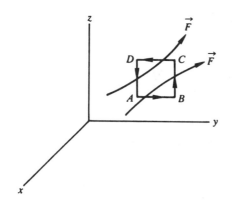

$$\text{curl}_x \, \mathbf{F} = \frac{F_y \, d_y + [F_z + (\partial F_z/\partial y)dy]dz - [F_y + (\partial F_y/\partial z)dz]dy - F_z \, dz}{dy \, dz}$$

$$= \frac{\partial F_z}{\partial y} - \frac{\partial F_y}{\partial z} \qquad (16\text{-}13)$$

Integration around the similar rectangular loops in the other two planes will yield $(\nabla \times F)_y$ and $(\nabla \times F)_z$.

Gravitational and electrostatic force fields are conservative. They can be expressed as gradients of scalar potentials and their curl is everywhere zero. The electric field produced by a changing magnetic field, however, is not

DIFFERENTIAL OPERATORS

conservative, nor is the magnetic field produced by an electric current or a changing electric field. Consider the force field represented by the diagram. It might be a map of magnetic field lines around a current-carrying wire going through the axis.

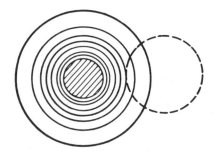

A line integral following any of the force lines does not cancel out to zero. A closed loop not containing the source current (such as the one shown by the dashed line) would yield a zero integral.

16.4 THE DEL OPERATOR IN CYLINDRICAL AND SPHERICAL COORDINATES.

In *cylindrical coordinates*:

$$\nabla\Phi = \hat{r}\frac{\partial\Phi}{\partial r} + \hat{\phi}\frac{1}{r}\frac{\partial\Phi}{\partial\phi} + \hat{k}\frac{\partial\Phi}{\partial z} \tag{16-14}$$

$$\nabla\cdot\mathbf{F} = \frac{1}{r}\frac{\partial(rF_r)}{\partial r} + \frac{1}{r}\frac{\partial F_\phi}{\partial\varphi} + \frac{\partial F_z}{\partial z} \tag{16-15}$$

$$\nabla\times\mathbf{F} = \hat{r}\frac{1}{r}\left[\frac{\partial F_z}{\partial\phi} - \frac{\partial(rF_\phi)}{\partial z}\right] + \hat{\phi}\left[\frac{\partial F_r}{\partial z} - \frac{\partial F_z}{\partial r}\right] + \hat{k}\frac{1}{r}\left[\frac{\partial(rF_\phi)}{\partial r} - \frac{\partial F_r}{\partial\phi}\right] \tag{16-16}$$

In *spherical coordinates*:

$$\nabla\phi = \hat{r}\frac{\partial\Phi}{\partial r} + \hat{\theta}\frac{1}{r}\frac{\partial\Phi}{\partial\theta} + \hat{\phi}\frac{1}{r\sin\theta}\frac{\partial\Phi}{\partial\phi} \tag{16-17}$$

$$\nabla\cdot\mathbf{F} = \frac{1}{r^2}\frac{\partial(r^2F_r)}{\partial r} + \frac{1}{r\sin\theta}\frac{\partial(\sin\theta F_\theta)}{\partial\theta} + \frac{1}{r\sin\theta}\frac{\partial F_\phi}{\partial\phi} \tag{16-18}$$

$$\nabla\times\mathbf{F} = \hat{r}\frac{1}{r\sin\theta}\left[\frac{\partial(\sin\theta F_\phi)}{\partial\theta} - \frac{\partial F_\theta}{\partial\phi}\right] + \hat{\theta}\frac{1}{r\sin\theta}\left[\frac{\partial F_r}{\partial\phi} - \frac{\partial(r\sin\theta F_\phi)}{\partial r}\right]$$
$$+ \hat{\phi}\frac{1}{r}\left[\frac{\partial(rF_\theta)}{\partial r} - \frac{\partial F_r}{\partial\theta}\right] \tag{16-19}$$

16.5 IDENTITIES CONCERNING THE DEL OPERATOR. The divergence of a gradient of a scalar function is usually written

$$\nabla^2 = \nabla \cdot \nabla = \frac{\partial^2}{\partial x^2} + \frac{\partial^2}{\partial y^2} + \frac{\partial^2}{\partial z^2} \qquad (16\text{--}20)$$

The divergence of the curl of a vector is necessarily zero, since

$$\nabla \cdot (\nabla \times \mathbf{F}) = \nabla \cdot \begin{vmatrix} \hat{i} & \hat{j} & \hat{k} \\ \dfrac{\partial}{\partial x} & \dfrac{\partial}{\partial y} & \dfrac{\partial}{\partial z} \\ F_x & F_y & F_z \end{vmatrix}$$

$$= \nabla \cdot \left[\hat{i}\left(\frac{\partial F_z}{\partial y} - \frac{\partial F_y}{\partial z}\right) + \hat{j}\left(\frac{\partial F_x}{\partial z} - \frac{\partial F_z}{\partial x}\right) + \hat{k}\left(\frac{\partial F_y}{\partial x} - \frac{\partial F_x}{\partial y}\right) \right]$$

$$= \frac{\partial}{\partial x}\left(\frac{\partial F_z}{\partial y} - \frac{\partial F_y}{\partial z}\right) + \frac{\partial}{\partial y}\left(\frac{\partial F_x}{\partial z} - \frac{\partial F_z}{\partial x}\right) + \frac{\partial}{\partial z}\left(\frac{\partial F_y}{\partial x} - \frac{\partial F_x}{\partial y}\right)$$

$$= \frac{\partial^2 F_z}{\partial x\,\partial y} - \frac{\partial^2 F_z}{\partial y\,\partial x} - \frac{\partial^2 F_y}{\partial x\,\partial z} + \frac{\partial^2 F_y}{\partial z\,\partial x} + \frac{\partial^2 F_x}{\partial y\,\partial z} - \frac{\partial^2 F_x}{\partial z\,\partial y} = 0$$

$$(16\text{--}21)$$

Therefore,

$$\nabla \cdot (\nabla \times \mathbf{F}) \equiv 0$$

The curl of the gradient of a scalar function is necessarily zero, since

$$\nabla \times \nabla \Phi = \nabla \times \left(\hat{i}\frac{\partial \Phi}{\partial x} + \hat{j}\frac{\partial \Phi}{\partial y} + \hat{k}\frac{\partial \Phi}{\partial z} \right)$$

$$= \begin{vmatrix} \hat{i} & \hat{j} & \hat{k} \\ \dfrac{\partial}{\partial x} & \dfrac{\partial}{\partial y} & \dfrac{\partial}{\partial z} \\ \dfrac{\partial \Phi}{\partial x} & \dfrac{\partial \Phi}{\partial y} & \dfrac{\partial \Phi}{\partial z} \end{vmatrix}$$

$$= \hat{i}\left| \frac{\partial^2 \Phi}{\partial y\,\partial z} - \frac{\partial^2 \Phi}{\partial x\,\partial z} \right| + \hat{j}\left| \frac{\partial^2 \Phi}{\partial x\,\partial z} - \frac{\partial^2 \Phi}{\partial x\,\partial z} \right|$$

$$+ \hat{k}\left| \frac{\partial^2 \Phi}{\partial x\,\partial z} - \frac{\partial^2 \Phi}{\partial x\,\partial z} \right| \equiv 0$$

Therefore,

$$\nabla \times \nabla \Phi \equiv 0 \qquad (16\text{--}22)$$

Two other useful identities that are related are

$$\text{grad div } \mathbf{F} = \nabla(\nabla \cdot \mathbf{F}) = \nabla^2 \mathbf{F} + \nabla \times (\nabla \times \mathbf{F}) \qquad (16\text{--}23)$$

$$\text{curl curl } \mathbf{F} = \nabla \times (\nabla \times \mathbf{F}) = \nabla(\nabla \cdot \mathbf{F}) - \nabla^2 \mathbf{F} \qquad (16\text{--}24)$$

16.6 AN APPLICATION OF THE DEL OPERATOR. As an example of the use of these general relationships, consider the derivation of a wave equation from Maxwell's equations for electromagnetism. In free space

(1) $\nabla \cdot E = 0$ $\qquad$ (2) $\nabla \cdot B = 0$

(3) $\nabla \times E = -\dfrac{1}{c} \dfrac{\partial B}{\partial t}$ $\qquad$ (4) $\nabla \times B = \dfrac{1}{c} \dfrac{\partial E}{\partial t}$

Take the curl of (3) and the time derivative of (4).

$$\nabla \times (\nabla \times E) = -\frac{1}{c} \nabla \times \frac{\partial B}{\partial t} \qquad \frac{\partial}{\partial t}(\nabla \times B) = \frac{1}{c} \frac{\partial^2 E}{\partial t^2}$$

Since

$$\nabla \times \frac{\partial B}{\partial t} = \frac{\partial}{\partial t}(\nabla \times B) \qquad \text{and} \qquad \nabla \times (\nabla \times E) = \nabla(\nabla \cdot E) - \nabla^2 E$$

$$\nabla(\nabla \cdot E) - \nabla^2 E = -\frac{1}{c^2} \frac{\partial^2 E}{\partial t^2}$$

Since $\nabla \cdot E = 0$,

$$\nabla^2 E - \frac{1}{c^2} \frac{\partial^2 E}{\partial t^2} = 0 \qquad\qquad (16\text{--}25)$$

This is a wave equation for the electric field vector, **E**. The same equation could have been derived much more laboriously by combining the detailed differentials corresponding to the divergence and curl.

APPENDICES

1. THE INTERNATIONAL SYSTEM OF UNITS (SI)

Sixty years after Professor Giorgi first proposed the basic system, the eleventh General Conference (1960) of the International Electrotechnical Commission and the International Union of Pure and Applied Physics adopted the SI physical units. The fundamental units are the meter (length), kilogram (mass), second (time), ampere (electric current), Kelvin degree (thermodynamic temperature), candela (luminous intensity), and mole (amount of substance). A crucial aspect of this system, and the one that caused the most controversy before the final adoption, is that B (magnetic flux) and H (magnetic field) must be quantities with different dimensions. The two are related by the equation, $B = \mu H$. In the c.g.s. system $\mu_0 = 1$ for free space, and in the e.m.u. system $\mu_0 = 1/c^2$. It is a pure number in both cases. In the SI units μ_0, the permeability of free space, has the dimensions $MLT^{-2}A^{-2}$ or MLQ^{-2}, and the value $4\pi \times 10^{-7}$ N/A^2. The compensating advantage of the SI units is that the practical units of volt and ampere become standard.

Besides the substantive problems involved in choosing the primacy of various electrical and mechanical quantities, the conference also faced the problem of adopting standard nomenclature. In general the symbols for units are written in lower case letters, except for the first letter when the name of a unit comes from a proper name. They should be the same for both singular and plural, and no period should follow them. For a compound unit formed by multiplication of two or more units, the symbol consists of the separate symbols with a raised dot between them—e.g., N·m for newton meter. Meter per second can be written either m/s or m·s^{-1}.

The following tables give the accepted names and symbols for the basic units and a variety of derived units.

SI units

Quantity	Name	Symbol	In terms of other units	In terms of SI fundamental units
Length	Meter	m		
Mass	Kilogram	kg		
Time	Second	s		
Electric current	Ampere	A		
Thermodynamic temperature	Kelvin	K		
Luminous intensity	Candela	cd		
Amount of substance	Mole	mol		
Plane angle	Radian	rad		
Solid angle	Steradian	sr		
Frequency	Hertz	Hz		s^{-1}
Force	Newton	N		$m \cdot kg \cdot s^{-2}$
Pressure	Pascal	Pa	N/m^2	$m^{-1} \cdot kg \cdot s^{-2}$
Energy or work	Joule	J	$N \cdot m$	$m^2 \cdot kg \cdot s^{-2}$
Power or radiant flux	Watt	W	J/s	$m^2 \cdot kg \cdot s^{-3}$
Electric charge	Coulomb	C	$A \cdot s$	$s \cdot A$
Potential difference	Volt	V	W/A	$m^2 \cdot kg \cdot s^{-3} \cdot A^{-1}$
Capacitance	Farad	F	C/V	$m^{-2} \cdot kg^{-1} \cdot s^2 \cdot A^2$
Electric resistance	Ohm	Ω	V/A	$m^2 \cdot kg \cdot s^{-3} \cdot A^{-2}$
Conductance	Siemens	S	A/V	$m^{-2} \cdot kg^{-1} \cdot s^3 \cdot A^2$
Magnetic flux	Weber	Wb	$V \cdot s$	$m^2 \cdot kg \cdot s^{-2} \cdot A^{-1}$
Magnetic flux density	Tesla	T	Wb/m^2	$kg \cdot s^{-2} \cdot A^{-1}$
Inductance	Henry	H	Wb/A	$m^2 \cdot kg \cdot s^{-2} \cdot A^{-2}$
Luminous flux	Lumen	lm		$cd \cdot sr$
Illuminance	Lux	lx		$m^{-2} \cdot cd \cdot sr$

SI prefixes

Factor	Prefix	Symbol	Factor	Prefix	Symbol
10^{12}	Tera	T	10^{-1}	Deci	d
10^{9}	Giga	G	10^{-2}	Centi	c
10^{6}	Mega	M	10^{-3}	Milli	m
10^{3}	Kilo	k	10^{-6}	Micro	μ
10^{2}	Hecto	h	10^{-9}	Nano	n
10^{1}	Deka	da	10^{-12}	Pico	p
			10^{-15}	Femto	f
			10^{-18}	Atto	a

2. PHYSICAL CONSTANTS*

The most accurate values of the fundamental physical constants are derived from critical analyses of all precision data that give either the value of a constant directly, or the value of a combination of such constants, such as the ratio of charge to mass of an electron.

The following table presents the most recent, accepted set of consistent values, as recommended by the Committee on Fundamental Constants of the National Academy of Sciences-National Research Council. All data are based on the scale of atomic masses in which the C^{12} isotope is assigned the mass of 12 u exactly. The $(\pm)$ values give the uncertainties in the last significant figure, determined in such a way that it is unlikely that the true value of this figure differs from the value given by more than the stated uncertainty.

The first three items in the table are purely *definitions*, included here for reference.

Standard gravitational acceleration (g_s)	$9.806\ 65$ m/s^2
Standard atmosphere (atm)	$101\ 325$ N/m^2
Thermochemical kilocalorie	4184 J
Speed of light in vacuum (c)	$2.997\ 925(\pm 3) \times 10^8$ m/s
Electronic charge (e)	$1.602\ 10(\pm 7) \times 10^{-19}$ C
Avogadro constant (N_A)	$6.0225(\pm 3) \times 10^{26}$/kmol
Faraday constant (F)	$9.6487(\pm 2) \times 10^7$ C/kmol
Planck constant (h)	$6.6256(\pm 5) \times 10^{-34}$ J$\cdot$s
	$= 4.1356 \times 10^{-15}$ eV$\cdot$s
Bohr magneton $(\mu_B = eh/4\pi m_e)$	$9.2732(\pm 6) \times 10^{-24}$ A$\cdot$m^2
Nuclear magneton $(\mu_N = eh/4\pi m_p)$	$5.0505(\pm 4) \times 10^{-27}$ A$\cdot$m^2
Universal gas constant (R)	$8314(\pm 1)$ J/kmol$\cdot$K
	$= 1.987$ kcal/kmol$\cdot$K
Volume per kilomole of ideal gas at 1 atm and $0°$C	$22.414(\pm 3)$ m^3/kmol
Boltzmann constant (k)	$1.3805(\pm 2) \times 10^{-23}$ J/K
Wien displacement constant (A)	$2.8978(\pm 4) \times 10^{-3}$ m$\cdot$K
Stefan–Boltzmann constant (σ)	$5.670(\pm 3) \times 10^{-8}$ W/K^4$\cdot$m^2
Gravitation constant (G)	$6.67(\pm 2) \times 10^{-11}$ N$\cdot$m^2/kg^2
Nuclidic mass unit (u)	$1.6604(\pm 1) \times 10^{-27}$ kg
Rest energy of one atomic mass unit	$931.48(\pm 2)$ MeV
Electron-volt (eV)	$1.602\ 10(\pm 7) \times 10^{-19}$ J

Rest masses of particles

	u	kg	MeV
Electron	$5.485\ 97(\pm 9) \times 10^{-4}$	$9.1091(\pm 4) \times 10^{-31}$	$0.511\ 006(\pm 5)$
Proton	$1.007\ 2766(\pm 2)$	$1.672\ 52(\pm 8) \times 10^{-27}$	$938.26(\pm 2)$
Neutron	$1.008\ 665(\pm 1)$	$1.674\ 82(\pm 8) \times 10^{-27}$	$939.55(\pm 2)$
Deuteron	$2.013\ 553$	3.3433×10^{-27}	1875.58
α-particle	$4.001\ 507$	6.6441×10^{-27}	3727.3

*From *Elements of Physics*, 5th Edition, George Shortley and Dudley Williams.

Astronomical data

Earth

Radius: mean	6371 km = 3959 mi
equatorial	6378 km = 3963 mi
polar	6357 km = 3950 mi
Distance from sun: mean	149.5×10^6 km = 92.9×10^6 mi
aphelion	152.1×10^6 km = 94.5×10^6 mi
perihelion	147.1×10^6 km = 91.4×10^6 mi
Period of rotation	86 164 s = 1 sidereal day = 23.94 h
Radiation from sun at earth's mean distance	1.35 kW/m^2

Moon

Radius	1 741 km = 1 082 mi
Distance from earth: mean	384 400 km = 239 000 mi
apogee	407 000 km = 253 000 mi
perigee	357 000 km = 222 000 mi
Period of revolution = period of rotation	27.322 d
Mass	7.343×10^{22} kg
Mean density	3.33 Mg/m^3

Sun

Radius	696 500 km = 432 200 mi
Mass	1.987×10^{30} kg
Mean density	1.41 Mg/m^3

Planets	Distance from sun (10^6 km)			Period of revolution (d)	Mean radius (km)	Mass[a] (10^{24} kg)	Mean density (Mg/m^3)
	Mean	Aphe-lion	Perihe-lion				
Mercury	57.9	69.8	46.0	88.0	2 420	3.167	5.46
Venus	108.1	109.0	107.5	224.7	6 261	4.870	4.96
Earth	149.5	152.1	147.1	365.2	6 371	5.975	5.52
Mars	227.8	249.2	206.6	687.0	3 389	0.639	4.12
Jupiter	777.8	815.9	740.7	4 333	69 900	1900	1.33
Saturn	1426	1508	1348	10 760	57 500	568.9	0.71
Uranus	2868	3007	2737	30 690	23 700	86.9	1.56
Neptune	4494	4537	4459	60 190	21 500	102.9	2.47
Pluto	5908	7370	4450	90 740	2 900	5.37	5.50

[a]Excluding satellites.

2. PHYSICAL CONSTANTS

3. CONVERSION FACTORS*

Angle	°	′	″	rad	rev
1 degree=	1	60	3600	1.745×10^{-2}	2.778×10^{-3}
1 minute=	1.667×10^{-2}	1	60	2.909×10^{-4}	4.630×10^{-5}
1 second=	2.778×10^{-4}	1.667×10^{-2}	1	4.848×10^{-6}	7.716×10^{-7}
1 radian=	57.30	3438	2.063×10^5	1	0.1592
1 revolution=	360	2.16×10^4	1.296×10^6	6.283	1

1 artillery mil=1/6400 rev=0.000 981 7 rad=0°.056 25

Length	m	km	i	f	mi
1 meter=	1	10^{-3}	39.37	3.281	6.214×10^{-4}
1 kilometer=	1000	1	3.937×10^4	3281	0.6214
1 inch=	0.0254	2.54×10^{-5}	1	0.0833	1.578×10^{-5}
1 foot=	0.3048	3.048×10^{-4}	12	1	1.894×10^{-4}
1 statute mile=	1609	1.609	6.336×10^4	5280	1

1 angstrom=10^{-10} m   1 millimicron (mμ)=1 nm   1 fathom=6 f
1 X-unit=10^{-13} m   1 light-year=9.4600×10^{12} km   1 yard (yd)=3 f
1 micron (μ)=1 μm   1 parsec=3.084×10^{13} km   1 rod=16.5 f
1 nautical mile=1853.2 m=1.1516 mi=6080.2 f   1 mil=10^{-3} i
1 astronomical unit=149.5×10^6 km   1 fermi=10^{-15} m   1 league=3 naut miles
1 Bohr radius=5.291 67$\times 10^{-11}$ m

Area	m^2	cm^2	f^2	i^2
1 square meter=	1	10^4	10.76	1550
1 square centimeter=	10^{-4}	1	1.076×10^{-3}	0.1550
1 square foot=	9.290×10^{-2}	929.0	1	144
1 square inch=	6.452×10^{-4}	6.452	6.944×10^{-3}	1

1 square mile=27 878 400 f^2=640 acre   1 acre=43 560 f^2 1 hectare=10 000 m^2
1 circular mil=7.854×10^{-7} i^2 1 barn=10^{-28} m^2 =2.471 acre

Volume	m^3	cm^3	f^3	i^3
1 cubic meter=	1	10^6	35.31	6.102×10^4
1 cubic centimeter=	10^{-6}	1	3.531×10^{-5}	0.06102
1 cubic foot=	2.832×10^{-2}	28,320	1	1728
1 cubic inch=	1.639×10^{-5}	16.39	5.787×10^{-4}	1

1 U.S. fluid gallon=4 quarts=8 pints=128 fluid ounces=231 i^3
1 British Imperial gallon=the volume of 10 lb of water at 62°F=277.42 i^3
1 liter=1000 cm^3

*From *Elements of Physics*, 5th Edition, George Shortley and Dudley Williams.

APPENDICES

Mass	g	kg	lb	sl	ton
1 gram=	1	0.001	0.002 205	6.852×10^{-5}	1.102×10^{-6}
1 kilogram=	1000	1	2.205	6.852×10^{-2}	1.102×10^{-3}
1 pound (avoirdupois)=	453.6	0.4536	1	3.108×10^{-2}	0.0005
1 slug=	1.459×10^4	14.59	32.17	1	1.609×10^{-2}
1 ton-mass=	9.072×10^5	907.2	2000	62.16	1

1 avoirdupois pound=16 avoirdupois ounces=7000 grains=0.453 592 37 kg
1 troy or apothecaries' pound=12 troy or apothecaries' ounces
 =0.8229 avoirdupois pound=5760 grains
1 long ton=2240 lb=20 cwt   1 stone=14 lb   1 hundredweight (cwt)=112 lb
1 metric ton=1000 kg=2205 lb   1 carat=0.2g   1 pennyweight (dwt)=24 grains
1 nuclidic mass unit (u)=1.6604×10^{-27} kg   1 quintal=100 kg

Time	y	d	h	min	s
1 year=	1	365.2	8.766×10^3	5.259×10^5	3.156×10^7
1 day=	2.738×10^{-3}	1	24	1440	86 400
1 hour=	1.141×10^{-4}	4.167×10^{-2}	1	60	3600
1 minute=	1.901×10^{-6}	6.944×10^{-4}	1.667×10^{-2}	1	60
1 second=	3.169×10^{-8}	1.157×10^{-5}	2.778×10^{-4}	1.667×10^{-2}	1

1 sidereal day=period of rotation of earth=86 164 s
1 year=period of revolution of earth=365.242 198 79 d

Density	sl/f^3	lb/f^3	lb/i^3	kg/m^3	g/cm^3
1 slug per f³=	1	32.17	1.862×10^{-2}	515.4	0.5154
1 pound per f³=	3.108×10^{-2}	1	5.787×10^{-4}	16.02	1.602×10^{-2}
1 pound per i³=	53.71	1728	1	2.768×10^4	27.68
1 kg per m³=	1.940×10^{-3}	6.243×10^{-2}	3.613×10^{-5}	1	0.001
1 gram per cm³=	1.940	62.43	3.613×10^{-2}	1000	1

Speed	f/s	km/h	m/s	mi/h	$knot$
1 foot per second=	1	1.097	0.3048	0.6818	0.5925
1 kilometer per hour=	0.9113	1	0.2778	0.6214	0.5400
1 meter per second=	3.281	3.6	1	2.237	1.944
1 mile per hour=	1.467	1.609	0.4470	1	0.8689
1 knot=	1.689	1.853	0.5148	1.152	1

1 knot=1 nautical mile/hr

Acceleration: 1 gal = 1 cm/s²

Force	dyne	kgf	N	p	pdl
1 dyne =	1	1.020×10^{-6}	10^{-5}	2.248×10^{-6}	7.233×10^{-5}
1 kilogram-force =	9.807×10^5	1	9.807	2.205	70.93
1 newton =	10^5	0.1020	1	0.2248	7.233
1 pound =	4.448×10^5	0.4536	4.448	1	32.17
1 poundal =	1.383×10^4	1.410×10^{-2}	0.1383	3.108×10^{-2}	1

1 kgf = 9.806 65 N   1 p = 32.173 98 pdl

Pressure	atm	inch of water	cm Hg	N/m²	p/i²
1 atmosphere =	1	406.8	76	1.013×10^5	14.70
1 inch of water^a =	2.458×10^{-3}	1	0.1868	249.1	0.03613
1 cm mercury^a =	1.316×10^{-2}	5.353	1	1333	0.1934
1 newton per m² =	9.869×10^{-6}	0.004 105	7.501×10^{-4}	1	1.450×10^{-4}
1 pound per i² =	6.805×10^{-2}	27.68	5.171	6.895×10^3	1

^{a}Under standard gravitational acceleration, and temperature of 4°C for water, 0°C for mercury.

1 bar = 10^5 N/m²   1 torr = 1 mm Hg   1 pascal = 1 N/m²
1 cm of water = 98.07 N/m²   1 f of water = 62.43 p/f²

Energy	BTU	fp	J	kcal	kWh
1 British thermal unit (BTU) =	1	777.9	1055	0.2520	2.930×10^{-4}
1 foot-pound =	1.285×10^{-3}	1	1.356	3.240×10^{-4}	3.766×10^{-7}
1 joule =	9.481×10^{-4}	0.7376	1	2.390×10^{-4}	2.778×10^{-7}
1 kilocalorie =	3.968	3086	4184	1	1.163×10^{-3}
1 kilowatt-hour =	3413	2.655×10^6	3.6×10^6	860.2	1

See also table of relativistic mass-energy equivalents on p. xi.
1 kcal = 2.612×10^{22} eV   1 horsepower-hour = 1.980×10^6 fp
1 erg = 10^{-7} joule   1 therm = 10^5 BTU
1 rydberg = 13.61 eV = 2.180×10^{-18} J

Power	BTU/h	fp/s	hp	kcal/s	kW	W
1 BTU/h =	1	0.2161	3.929×10^{-4}	7.000×10^{-5}	2.930×10^{-4}	0.2930
1 fp/s =	4.628	1	1.818×10^{-3}	3.239×10^{-4}	1.356×10^{-3}	1.356
1 horsepower =	2545	550	1	0.1782	0.7457	745.7
1 kcal/s =	1.429×10^4	3087	5.613	1	4.184	4184
1 kilowatt =	3413	737.6	1.341	0.2390	1	1000
1 watt =	3.413	0.7376	1.341×10^{-3}	2.390×10^{-4}	0.001	1

1 ton (refrigeration) = 12 000 BTU/h

Electric charge	abC	C	$statC$
1 abcoulomb (1 EMU) =	1	10	2.998×10^{10}
1 coulomb =	0.1	1	2.998×10^{9}
1 statcoulomb (1 ESU) =	3.336×10^{-11}	3.336×10^{-10}	1

1 franklin = 1 Fr = 1 statC   1 ampere-hour = 3600 C

Electric current	abA	A	$statA$
1 abampere (1 EMU) =	1	10	2.998×10^{10}
1 ampere =	0.1	1	2.998×10^{9}
1 statampere (1 ESU) =	3.336×10^{-11}	3.336×10^{-10}	1

1 biot = 1 Bi = 1 abA

Electric potential	abV	V	$statV$
1 abvolt (1 EMU) =	1	10^{-8}	3.336×10^{-11}
1 volt =	10^{8}	1	3.336×10^{-3}
1 statvolt (1 ESU) =	2.998×10^{10}	299.8	1

Electric resistance	$abohm$	Ω	$statohm$
1 abohm (1 EMU) =	1	10^{-9}	1.113×10^{-21}
1 ohm =	10^{9}	1	1.113×10^{-12}
1 statohm (1 ESU) =	8.987×10^{20}	8.987×10^{11}	1

Capacitance	abF	F	μF	$statF$
1 abfarad (1 EMU) =	1	10^{9}	10^{15}	8.987×10^{20}
1 farad =	10^{-9}	1	10^{6}	8.987×10^{11}
1 microfarad =	10^{-15}	10^{-6}	1	8.987×10^{5}
1 statfarad (1 ESU) =	1.113×10^{-21}	1.113×10^{-12}	1.113×10^{-6}	1

Inductance	abH	H	mH	$statH$
1 abhenry (1 EMU) =	1	10^{-9}	10^{-6}	1.113×10^{-21}
1 henry =	10^{9}	1	1000	1.113×10^{-12}
1 millihenry =	10^{6}	0.001	1	1.113×10^{-15}
1 stathenry (1 ESU) =	8.987×10^{20}	8.987×10^{11}	8.987×10^{14}	1

Magnetic flux	Mx	kiloline	Wb
1 maxwell (1 line or 1 EMU) =	1	0.001	10^{-8}
1 kiloline =	1000	1	10^{-5}
1 weber =	10^8	10^5	1

1 ESU = 299.8 weber

Magnetic intensity $\mathcal{B}$	G	kiloline/i^2	T	mG
1 gauss (line per cm²) =	1	6.452×10^{-3}	10^{-4}	1000
1 kiloline per square inch =	155.0	1	1.550×10^{-2}	1.550×10^5
1 tesla =	10^4	64.52	1	10^7
1 milligauss =	0.001	6.452×10^{-6}	10^{-7}	1

1 T = 1 Wb/m²   1 ESU = 2.998×10^6 T   1 gamma = 10^{-2} mG = 10^{-9} T

Magnetomotive force	abA-turn	A-turn	Gi
1 abampere-turn =	1	10	12.57
1 ampere-turn =	0.1	1	1.257
1 gilbert =	7.958×10^{-2}	0.7958	1

1 ESU = 1 statampere-turn = 3.336×10^{-10} A-turn

Magnetizing force $\mathcal{K}$	abA/cm	A/in	A/m	Oe
1 abampere-turn per centimeter =	1	25.40	1000	12.57
1 ampere-turn per inch =	3.937×10^{-2}	1	39.37	0.4947
1 ampere-turn per meter =	0.001	2.540×10^{-2}	1	1.257×10^{-2}
1 oersted =	7.958×10^{-2}	2.021	79.58	1

1 oersted = 1 gilbert/cm   1 ESU = 3.336×10^{-8} A-turn/m

Mass-energy equivalents	kg	u	J	MeV
1 kilogram ~	1	6.025×10^{26}	8.987×10^{-16}	5.610×10^{29}
1 nuclidic mass unit ~	1.660×10^{-27}	1	1.492×10^{-10}	931.5
1 joule ~	1.113×10^{-17}	6.705×10^9	1	6.242×10^{12}
1 million electron-volts ~	1.783×10^{-30}	1.074×10^{-3}	1.602×10^{-13}	1

4. USED FORMULAS

Kinematics and dynamics

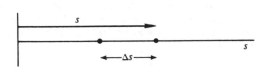 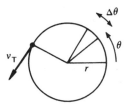

Linear motion   Rotational Motion

$$\bar{v} = \frac{\Delta s}{\Delta t} \quad v = \frac{ds}{dt} \qquad\qquad \bar{\omega} = \frac{\Delta \theta}{\Delta t} \quad \omega = \frac{d\theta}{dt}$$

$$\bar{a} = \frac{\Delta v}{\Delta t} \quad a = \frac{dv}{dt} = \frac{d^2s}{dt^2} \qquad \bar{\alpha} = \frac{\Delta \omega}{\Delta t} \quad \alpha = \frac{d\omega}{dt} = \frac{d^2\theta}{dt^2}$$

$$v_{\text{Tangential}} = \omega r$$
$$a_{\text{Tangential}} = \alpha r$$

For Constant Acceleration:
$$\Delta s = v_0 \Delta t + \tfrac{1}{2} a (\Delta t)^2 \qquad\qquad \Delta \theta = \omega_0 \Delta t + \tfrac{1}{2} \alpha (\Delta t)^2$$
$$v_f^2 = v_0^2 + 2a\Delta s \qquad\qquad\qquad \omega_f^2 = \omega_0^2 + 2\alpha \Delta \theta$$

Torque: $\vec{\tau} = \vec{r} \times \vec{F}$

Moment of Inertia: $I = \sum_i m_i r_i^2 = \int_V r^2 dm$

Angular Momentum: $\vec{L} = \vec{r} \times \vec{p} = \vec{r} \times \overrightarrow{mv}$

Newton's Laws:

   I. $a = 0$ if $F = 0$ $\alpha = 0$ if $\tau = 0$
                                      (for constant I)

   II. $F = ma = \dfrac{d(mv)}{dt}$ $\tau = I\alpha = \dfrac{dL}{dt}$

   III. $\Delta(mv)_{\text{system}} = 0$ $\Delta L_{\text{system}} = 0$
           $F_{12} = -F_{21}$ $\tau_{12} = -\tau_{21}$

Kinetic Energy:
$$E_{\text{kin}} = \tfrac{1}{2} mv^2 \qquad E_{\text{kin}} = \tfrac{1}{2} I\omega^2$$

Relationship between Force and Potential Energy:
$$\vec{F} = -\frac{dV(r)}{dr} \qquad\qquad \Delta V_{A-B} = -\int_A^B \vec{F} \cdot \vec{dr}$$

Relationship between Force and Momentum:
$$F = \frac{d(mv)}{dt} \text{ or in general, } F = \frac{dp}{dt}$$

Gravitation:
$$F = G\frac{m_1 m_2}{r^2}$$
For objects with spherical mass distribution and further apart than the sum of their radii.
$$E_{\text{grav}} = -G\frac{m_1 m_2}{r}$$
For spheres, r is distance between centers.

Circular Motion:
$$a_{\text{centripetal}} = \frac{v^2}{r} = \omega^2 r = 4\pi^2 f^2 r = \frac{4\pi^2}{T^2} r$$

Energy Stored in Spring:
$$V_{\text{pot}} = \tfrac{1}{2} kx^2$$
restoring force is
$$F = -kx$$

Total Mass-Energy:	$$E = mc^2 = \frac{m_0 c^2}{\sqrt{1 - \frac{v^2}{c^2}}}$$ $$= m_0 c^2 + \frac{1}{2} m_0 v^2 + \frac{3}{8} m_0 \frac{v^4}{c^2} + \cdots$$
Friction between Dry Sliding Objects:	It is approximately true that for many surfaces, $F_{\text{friction}} = \mu F_{\text{normal}}$, where μ depends only on the nature of the surfaces and is independent of velocity over a considerable range.
Conditions for Static Equilibrium:	$\Sigma \vec{F} = 0 \qquad \Sigma \vec{\tau} = 0$

Units for Mechanics:

MKS m in kilograms, s in meters, t in seconds, F in newtons

British m in pounds/g (lbs/32.2), s in feet, t in seconds, F in pounds

Simple Harmonic Motion—SHM:

$$x = A \sin(\omega t + \alpha) = A \sin\left(2\pi \frac{t}{T} + \alpha\right)$$
$$= A \sin(2\pi f t + \alpha)$$
$$v = -A\omega \cos(\omega t + \alpha)$$
$$a = -A\omega^2 \sin(\omega t + \alpha)$$

A is the amplitude of oscillation
ω is the angular frequency in radians/sec
f is the cyclical frequency in Hz (cycles/sec)
T is the period
α is the phase angle at $t = 0$

For small amplitudes

$$T = 2\pi\sqrt{\frac{m}{k}}$$ For mass, m, of bob on spring with constant, k.

$$T = 2\pi\sqrt{\frac{l}{g}}$$ For simple pendulum of length l, in gravitational field, g.

$$T = 2\pi\sqrt{\frac{I}{Mgd}}$$ For "physical" pendulum where I is moment of inertia around axis, M is mass of object, d is distance from axis to center of gravity, and g is gravitational field.

Hydrodynamics

Continuity in Fluid Flow:

$$\nabla \cdot (\overrightarrow{\rho v}) + \frac{d\rho}{dt} = Q$$

ρ is density, v is velocity Q is mass per unit time per unit volume from source

Bernoulli's:

$$p_1 + \tfrac{1}{2}\rho v_1^2 + \rho g h_1 = p_2 + \tfrac{1}{2}\rho v_2^2 + \rho g h_2$$

where p is pressure, ρ is density, v is velocity, g is gravitational field, h is height above arbitrary level. Subscripts 1 and 2 denote two different locations in pipe. Bernoulli's formula takes no account of friction and so is usually a poor approximation to real liquid flow.

Elasticity

Young's Modulus:
$$E = \frac{\frac{\Delta F}{A}}{\frac{\Delta l}{L}} = \frac{L}{A}\frac{\Delta F}{\Delta l}$$

Bulk Modulus:
$$B = -\frac{\Delta p}{\frac{\Delta V}{V}} = -V\frac{\Delta p}{\Delta V} \qquad \text{compressibility} = k = \frac{1}{B}$$

Torsion or Shear Modulus:
$$\frac{\frac{\Delta F}{A}}{\Delta \tan \beta} \approx \frac{\frac{\Delta F}{A}}{\Delta \beta} = \frac{1}{A}\frac{\Delta F}{\Delta \beta}$$

Wave Motion

$$f\lambda = v$$

f is frequency in *Hz*
λ is wavelength
v is velocity of wave (phase velocity)

Traveling Wave:
$y = f(x + vt)$ ⟵⋀⋀⋀ $y = f(x - vt)$ ⋀⋀⋀⟶
$$y = A \sin \frac{2\pi}{\lambda}(x \pm vt) = A \sin 2\pi\left(\frac{x}{\lambda} \pm \frac{t}{T}\right)$$
$$y = A \sin(kx \pm \omega t) \quad \text{or} \quad = Ae^{i(kx \pm \omega t)}$$
where $k = \frac{2\pi}{\lambda}$ is the "wave number"

and $\omega = 2\pi f = \frac{2\pi}{T}$ is the angular frequency in radians/sec

Standing Wave:
$$y = 2A \sin kx \cos \omega t$$

Wave Velocity:
$v = k\sqrt{\frac{F}{\mu}}$ where F is a tension and μ is inertial linear density

For small amplitudes on a rope, $v = \sqrt{\frac{F}{\mu}}$ where F is tension and μ is mass per unit length

In elastic medium, $v = \sqrt{\frac{B}{\rho_0}}$ where B is bulk modulus and ρ_0 is undisturbed density

For a gas, $v = \sqrt{\frac{\gamma P_0}{\rho_0}}$ where γ is the ratio of specific heats and P_0 is the undisturbed pressure.

Power Density Transmitted in a Wave:
$P = A^2\omega^2\mu v \sin^2(kx - \omega t)$
$\bar{P} = 2\pi^2 A^2 f^2 \mu v$ where μ is mass density

Doppler Effect with
Stationary Medium:

Observer in motion-
toward source

$$f' = f\left(1 + \frac{v_0}{v}\right) = f\left(\frac{v + v_0}{v}\right)$$

$$\lambda' = \lambda\left(\frac{v}{v + v_0}\right) = \lambda\left(\frac{1}{1 + \frac{v_0}{v}}\right)$$

Observer in motion-
away from source

$$f' = f\left(1 - \frac{v_0}{v}\right) = f\left(\frac{v - v_0}{v}\right)$$

$$\lambda' = \lambda\left(\frac{v}{v - v_0}\right) = \lambda\left(\frac{1}{1 - \frac{v_0}{v}}\right)$$

Source in motion-
toward observer

$$f' = f\left(\frac{v}{v - v_s}\right) = f\left(\frac{1}{1 - \frac{v_s}{v}}\right)$$

$$\lambda' = \lambda\left(\frac{v - v_s}{v}\right) = \lambda\left(T - \frac{v_s}{v}\right)$$

Source in motion-
away from observer

$$f' = f\left(\frac{v}{v + v_s}\right) = f\left(\frac{1}{1 + \frac{v_s}{v}}\right)$$

$$\lambda' = \lambda\left(\frac{v + v_s}{v}\right) = \lambda\left(1 + \frac{v_s}{v}\right)$$

The primed frequencies and wavelength are those
observed while the unprimed are those that are generated
or that would be observed if there were no motion of
observer and source. The subscript "0" denotes
"observer"; the subscript "s" denotes "source". The
wave velocity in the medium is v.

Doppler Effect for Light
in Vacuum:

Observer and source separating

$$f' = f\sqrt{\frac{1 - \frac{v}{c}}{1 + \frac{v}{c}}} = f\frac{1 - \frac{v}{c}}{\sqrt{1 - \frac{v^2}{c^2}}}$$

$$= f\left[1 - \frac{v}{c} + \frac{1}{2}\left(\frac{v}{c}\right)^2 + \cdots\right]$$

$$\lambda' = \lambda\sqrt{\frac{1 + \frac{v}{c}}{1 - \frac{v}{c}}} = \lambda\frac{1 + \frac{v}{c}}{\sqrt{1 - \frac{v^2}{c^2}}}$$

$$= \lambda\left[1 + \frac{v}{c} + \frac{1}{2}\left(\frac{v}{c}\right)^2 + \cdots\right]$$

Observer and source approaching

$$f' = f\sqrt{\frac{1 + \frac{v}{c}}{1 - \frac{v}{c}}} = f\frac{1 + \frac{v}{c}}{\sqrt{1 - \frac{v^2}{c^2}}}$$

$$= f\left[1 + \frac{v}{c} + \frac{1}{2}\left(\frac{v}{c}\right)^2 + \cdots\right]$$

$$\lambda' = \lambda\sqrt{\frac{1 - \frac{v}{c}}{1 + \frac{v}{c}}} = \lambda\frac{1 - \frac{v}{c}}{\sqrt{1 - \frac{v^2}{c^2}}}$$

$$= \lambda\left[1 - \frac{v}{c} + \frac{1}{2}\left(\frac{v}{c}\right)^2 + \cdots\right]$$

Thermodynamics

$$T_c = \tfrac{5}{9}(T_F - 32) = T_K - 273.15$$
(Celsius)   (Fahrenheit)   (Kelvin)

Q is symbol for heat. dQ is symbol for infinitesimal amount of heat. It is not an exact differential since the heat emitted or absorbed by a system in transition between two states depends on the path (whether at constant volume, constant pressure, etc.)

v is symbol for internal energy
W is symbol for external work performed
Q is usually measured in calories   (4.19 joules = 1 calorie).

Heat Capacity:

$$C = \frac{dQ}{dT}$$

Newton's Law of Heat Conduction:

$$\frac{dQ}{dt} = -kA\frac{dT}{dx}$$

The flow of heat is proportional to the area through which it is flowing and to the temperature gradient.

First Law of Thermodynamics:

$$dQ = dv + dW$$

for a gas at constant pressure: $dQ = C_p dT + P\,dV$
for a gas at constant volume: $dQ = C_v\,dT$
for a gas at constant temperature: $dQ = P\,dV$

Enthalpy:

$$H = v + PV$$
$$dH = dv + P\,dV + V\,dP = dQ + V\,dP$$

	Internal energy	Enthalpy
	$dv = dQ - P\,dV$	$dH = dQ + V\,dP$
constant V:	$dv = dQ = C_v\,dT$	$dH = dQ = C_p\,dT$
adiabatic:	$dv = -P\,dV$	$dH = V\,dP$
Free expansion:	$dv = 0$	throttling process: $dH = 0$

Equation of State for Ideal Gas:

$PV = nRT$ n is number of moles
$PV = NkT$ N is number of molecules
R is universal gas constant = 8.314 joule/mole (K)
                             = 1.986 cal/mole (K)

N_0 is Avogadro's number
= 6.023×10^{23} molecules/mole
k is Boltzmann's constant

$$= \frac{R}{N_0} = 1.38 \times 10^{-23} \text{ joule/molecule (K)}$$

Relationships between Specific Heats:

$C_p - C_v = R$ (for an ideal gas)
$C_p/C_v = \gamma$ ratio of specific heats
$PV^\gamma = K$ for ideal gas undergoing adiabatic change (without heat transfer)

Maxwellian Distribution of Velocities in a Gas:

$$N_v dv = 4\pi N\left(\frac{m}{2\pi kT}\right)^{3/2} v^2 e^{-(mv^2/2kT)}\,dv$$

Van der Waal's Equation for a Gas:

$$\left(P + \frac{n^2 a}{V^2}\right)(V - nb) = nRT$$

Efficiency of Carnot Cycle:

$$\text{efficiency} = \frac{Q_1 - Q_2}{Q_1} = \frac{T_1 - T_2}{T_1}$$

Reversible Cycle:	$\oint_R \dfrac{dQ}{T} = 0$

Entropy: $dS = \dfrac{dQ}{T}$ $S = k \ln w$ k is Boltzmann's constant
w is probability of state

Second Law of
Thermodynamics: $dS \geqq 0$ for entire system in any process

Optics

Snell's Law of Refraction: $\dfrac{\sin \theta_1}{\sin \theta_2} = n_{12} = \dfrac{v_1}{v_2} = \dfrac{n_2}{n_1}$

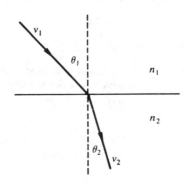

$$n = \frac{c}{v}$$

in general: $n_1 \sin \theta_1 = n_2 \sin \theta_2 = n_3 \sin \theta_3$, etc.

Critical angle of refraction: $\sin \theta_c = \dfrac{n_2}{n_1}$

Thin Lens Equation:

$$\frac{1}{o} + \frac{1}{i} = \frac{1}{f}$$

$$\frac{1}{o} + \frac{1}{i} = (n-1)\left(\frac{1}{R_1} + \frac{1}{R_2}\right)$$

Magnification: $m = \dfrac{o}{i}$

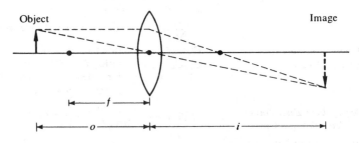

R_1 and R_2 are radii of curvature of lens surfaces

Maxima for Interference in Double Slit Image:	$d \sin \theta = m\lambda$ (m is any integer)

Minima of Diffraction from Single Slit:	$a \sin \theta = m\lambda$ (a is width of slit)
First Minimum for Circular Pinhole:	$\sin \theta = 1.22\dfrac{\lambda}{d}$ (d is diameter of pinhole)
Maxima from Grating:	$d \sin \theta = m\lambda$ (d is distance between slits)
Brewster's Angle for Completely Polarized Reflection:	$\tan \theta = n$
Black Body Radiancy— Stefan-Boltzmann Law:	$R = \sigma T^4$
Planck's Radiation Formula:	$E_\nu \, d\nu = \dfrac{8\pi h\nu^3}{c^3(e^{h\nu/kT} - 1)}d\nu$ $E_\lambda \, d\lambda = \dfrac{hc^2}{\lambda^5}\dfrac{8\pi}{e^{hc/kT\lambda} - 1}d\lambda$
Photon Energy:	$E = h\nu = \dfrac{hc}{\lambda}$ h is Planck's constant $=$ 6.63×10^{-34} joule·sec
Photon Momentum:	$p = \dfrac{E}{c} = \dfrac{h}{\lambda}$

Relativity

$$E = mc^2 \qquad E^2 = p^2c^2 + (m_0c^2)^2$$

$$m = \frac{m_0}{\sqrt{1 - \beta^2}} \quad \Delta t = \frac{\Delta t_0}{\sqrt{1 - \beta^2}} \quad \Delta l = \Delta l_0\sqrt{1 - \beta^2}$$

where $\beta = \dfrac{v}{c}$

Addition of Velocities:	$w = \dfrac{u + v}{1 + \dfrac{uv}{c^2}}$

Transformations:

Galilean:

$x = x_0 + vt$

$y = y_0$

$z = z_0$

$t = t_0$

Lorentz:

$x = \dfrac{x_0 + vt_0}{\sqrt{1 - \beta^2}}$

$y = y_0$

$z = z_0$

$t = \dfrac{t_0 + \dfrac{x_0 v}{c^2}}{\sqrt{1 - \beta^2}}$

4. USED FORMULAS

Electromagnetism

Coulomb's Law for Point Charges:

$$F = \frac{1}{4\pi\epsilon_0}\frac{q_1 q_2}{r^2} = 9 \times 10^9 \frac{q_1 q_2}{r^2}$$

q in coulombs
r in meters
F in newtons
$\epsilon_0 = 8.85 \times 10^{-12}$ coul^2/nt$\cdot$m^2

$$F = \frac{q_1 q_2}{r^2}$$

q in e.s.u.
r in centimeters
F in dynes

1 coulomb $= 3 \times 10^{10}$ e.s.u.

$$\boxed{\text{coulomb}} \qquad \square \text{ e.s.u.}$$

Electric Field:

$$\vec{E} = \frac{\vec{F}}{q_0}$$

E in newtons/coulomb or volts/meter

Electric Potential:

$$\Delta V = \frac{\Delta \text{ Work}}{q_0}$$

V in joules/coulomb or volts

$$\vec{E} = -\nabla V \qquad E_x = -\frac{dV}{dx}$$

$$\Delta V = \frac{\Delta \text{ Work}}{q_0}$$

If q_0 is in e.s.u. and work is in ergs, V is in statvolts.

1 statvolt $= 300$ volts

$$\boxed{\text{statvolt}} \qquad \square \text{ volt}$$

Flux:

$$\Phi_E = \oint_{\text{surface}} \vec{E} \cdot \vec{ds}$$

Gauss' Law:

$$\epsilon_0 \oint_s \vec{E} \cdot \vec{ds} = q \quad \text{in MKS units and coulombs}$$

$$\oint_s \vec{E} \cdot \vec{ds} = 4\pi q \quad \text{in c.g.s. units and e.s.u.}$$

Potential from Point Charge:

$$\Delta V_{r_1 - r_2} = \frac{q}{4\pi\epsilon_0}\left[\frac{1}{r_1} - \frac{1}{r_2}\right]$$

if $r_2 \longrightarrow \infty$, $\quad V_r = \frac{q}{4\pi\epsilon_0}\frac{1}{r}$

in c.g.s. and e.s.u. units, $\quad V_r = \frac{q}{r}$

Potential from Dipole:

$$V = \frac{1}{4\pi\epsilon_0}\frac{p\cos\theta}{r^2}$$

Capacitance:

$$C = \frac{q}{V}$$

C in farads
q in coulombs
V in volts

$$v = \frac{1}{2}CV^2 = \frac{1}{2}\frac{q^2}{C} = \frac{1}{2}qV$$

Energy stored in a condenser

parallel plate: $\quad C = \frac{\kappa \epsilon_0 A}{d}$

C in farads
κ is dielectric constant
$\epsilon_0 = 8.85 \times 10^{-12}$ coul²/nt·m² or farads/meter
A in meters squared
d is spacing in meters

In e.s.u. units: $\quad C = \frac{\kappa A}{4\pi d}$

C in centimeters
κ is dielectric constant
A in centimeters squared
d in centimeters
Capacity of air (vacuum) capacitor, 1 cm², with spacing of 0.1 cm:

$$C = \frac{1 \cdot 8.85 \times 10^{-12} \cdot 10^{-4}}{10^{-3}} \approx 1\mu\mu f \text{ or } 1 \text{ pf (picofarad)}$$

$$C = \frac{1 \cdot 1}{4\pi \cdot 10^{-1}} \approx 1 \text{ cm}$$

Current:

$i = \frac{dq}{dt} \quad$ current density: $\quad j = \frac{i}{A}$

1 ampere (coulomb/second) = 3×10^9 e.s.u./second

Ohm's Law:

$I = \dfrac{V}{R} \quad \begin{array}{l} I \text{ in amperes} \\ V \text{ in volts} \\ R \text{ in ohms} \end{array}$

Electrical Power:

$P = I^2 R = \dfrac{V^2}{R} = IV$

Lorentz Force:

$d\vec{F} = i\,d\vec{l} \times \vec{B}$
$\vec{F} = q\vec{E} + q\,\vec{v} \times \vec{B}$

Ampère's Law:

$\dfrac{1}{\mu_0} \displaystyle\oint_{\text{loop}} \vec{B} \cdot d\vec{l} = i \qquad \begin{array}{l} \mu_0 = 4\pi \times 10^{-7} \\ \text{weber/amp·m} \end{array}$

B is magnetic intensity in weber/m²
In electromagnetic units (e.m.u.):

$$\oint \vec{H} \cdot d\vec{l} = 4\pi i$$

H is magnetic field in gauss
i is measured in abamps; 1 abamp = 10 amperes
In vacuum (or air), 1 weber/m² = 10^4 gauss

Field near Long Wire:

$B = \dfrac{\mu_0}{2\pi} \dfrac{i}{r} \quad$ (r in meters, i in amps)

$H = \dfrac{2i}{r} \quad$ (r in centimeters, i in abamps)

Force between Two Wires:

$\dfrac{F}{l} = \dfrac{\mu_0 i_1 i_2}{2\pi d} \quad$ (newtons/meter)

$\dfrac{F}{l} = \dfrac{2i_1 i_2}{d} \quad$ (dynes/cm)

e.g. force between two wires, 1 m long, 1 cm apart, carrying 100 amperes:

$$F = \frac{4\pi \times 10^{-7} \cdot 10^4}{2\pi \cdot 10^{-2}} \cdot 1 = 2 \times 10^{-1} \text{ N}$$

$$F = \frac{2 \cdot 10^2}{1} \cdot 10^2 = 2 \times 10^4 \text{ dynes} = 2 \times 10^{-1} \text{ N}$$

Field Inside Long Solenoid:
$B = \mu_0 in$ n is number of turns/m
$H = 4\pi ni$ n is number of turns/cm

Law of Biot-Savart:
$$dB = \frac{\mu_0 i}{4\pi} \frac{\vec{dl} \times \vec{r}}{r^3}$$

Induced EMF:
(Faraday's Law)
$$V = \frac{d\Phi}{dt}$$ V is in volts
Φ is total magnetic flux in webers

$$V = 10^8 \frac{d(A \cdot H)}{dt}$$ V is in volts
H is in gauss
A is in cm^2

Inductive Reactance:
$$X_L = 2\pi f L$$ f is frequency of a.c.
L is inductance in henries
X_L is in ohms

Capacitive Reactance:
$$X_C = \frac{1}{2\pi f C}$$ C is in farads
X_C is in ohms

Impedance of Series Circuit:
$$Z = R + i\left(\omega L - \frac{1}{\omega C}\right) \qquad i = \sqrt{-1}$$

$$|Z| = \sqrt{R^2 + \left(\omega L - \frac{1}{\omega c}\right)^2}$$

Maxwell's Equations for Free Space:

SI Units   (MKS)

$$\nabla \cdot \vec{E} = \frac{\rho}{\epsilon_0} \qquad\qquad \oint_s \vec{E} \cdot \vec{dS} = \oint_v \frac{\rho}{\epsilon_0} dV$$
$$= \frac{q}{\epsilon_0} \text{ enclosed}$$

$$\nabla \cdot \vec{B} = 0 \qquad\qquad \oint_s \vec{B} \cdot \vec{ds} = 0$$

$$\nabla \times \vec{E} = -\frac{\partial \vec{B}}{\partial t} \qquad\qquad \oint_L \vec{E} \cdot \vec{dl} = -\frac{\partial}{\partial t} \iint_A \vec{B} \cdot \vec{dA}$$

$$c^2 \nabla \times \vec{B} = \frac{\vec{j}}{\epsilon_0} + \frac{\partial \vec{E}}{\partial t} \qquad c^2 \oint_L \vec{B} \cdot \vec{dl} = \iint_A \frac{\vec{j}}{\epsilon_0} \cdot \vec{dA}$$
$$+ \frac{\partial}{\partial t} \iint_A \vec{E} \cdot \vec{dA}$$

Gaussian Units (c.g.s. and e.s.u.)

$$\nabla \cdot \vec{E} = 4\pi\rho \qquad\qquad \oint_S \vec{E} \cdot \vec{ds} = \oint_V 4\pi\rho dV$$
$$= 4\pi q \text{ enclosed}$$

$$\nabla \cdot \vec{B} = 0 \qquad\qquad \oint_S \vec{B} \cdot \vec{dS} = 0$$

$$\nabla \times \vec{E} = -\frac{1}{c} \frac{\partial \vec{B}}{\partial t} \qquad\qquad \oint_L \vec{E} \cdot \vec{dl} = -\frac{1}{c} \frac{\partial}{\partial t} \iint_A \vec{B} \cdot \vec{dA}$$

$$\nabla \times \vec{B} = 4\pi \frac{\vec{j}}{c} + \frac{1}{c} \frac{\partial \vec{E}}{\partial t} \qquad \oint_L \vec{B} \cdot \vec{dl} = \iint_A 4\pi \frac{\vec{j}}{c} \cdot \vec{dA}$$
$$+ \frac{1}{c} \frac{\partial}{\partial t} \iint_A \vec{E} \cdot \vec{dA}$$

5. ERROR FUNCTION INTEGRAL

$$y(x) = \frac{2}{\sqrt{\pi}} \int_0^x e^{-t^2}\, dt$$

x	0	1	2	3	4	5	6	7	8	9
0.0	.0000	.0113	.0226	.0338	.0451	.0564	.0676	.0789	.0901	.1013
0.1	.1125	.1236	.1348	.1459	.1570	.1680	.1790	.1900	.2010	.2118
0.2	.2227	.2335	.2443	.2550	.2657	.2763	.2869	.2974	.3079	.3183
0.3	.3286	.3389	.3491	.3593	.3694	.3794	.3893	.3992	.4090	.4187
0.4	.4284	.4380	.4475	.4569	.4662	.4755	.4847	.4938	.5028	.5117
0.5	.5205	.5292	.5379	.5465	.5549	.5633	.5716	.5798	.5879	.5959
0.6	.6039	.6117	.6194	.6271	.6346	.6420	.6494	.6566	.6638	.6708
0.7	.6778	.6847	.6914	.6981	.7047	.7112	.7175	.7238	.7300	.7361
0.8	.7421	.7480	.7538	.7595	.7651	.7707	.7761	.7814	.7867	.7918
0.9	.7969	.8019	.8068	.8116	.8163	.8209	.8254	.8299	.8342	.8385
1.0	.8427	.8468	.8508	.8548	.8587	.8624	.8661	.8698	.8733	.8768
1.1	.8802	.8835	.8868	.8900	.8931	.8961	.8991	.9020	.9048	.9076
1.2	.9103	.9130	.9155	.9181	.9205	.9229	.9252	.9275	.9297	.9319
1.3	.9340	.9361	.9381	.9400	.9419	.9438	.9456	.9473	.9490	.9507
1.4	.9523	.9539	.9554	.9569	.9583	.9597	.9611	.9624	.9637	.9649
1.5	.9661	.9673	.9684	.9695	.9706	.9716	.9726	.9736	.9746	.9755
1.6	.9764	.9772	.9780	.9788	.9796	.9804	.9811	.9818	.9825	.9832
1.7	.9838	.9844	.9850	.9856	.9861	.9867	.9872	.9877	.9882	.9886
1.8	.9891	.9895	.9899	.9904	.9907	.9911	.9915	.9918	.9922	.9925
1.9	.9928	.9931	.9934	.9937	.9939	.9942	.9944	.9947	.9949	.9951
2.0	.9953	.9955	.9957	.9959	.9961	.9963	.9964	.9966	.9967	.9969
2.1	.9970	.9972	.9973	.9974	.9975	.9976	.9978	.9979	.9980	.9981
2.2	.9981	.9982	.9983	.9984	.9985	.9985	.9986	.9987	.9987	.9988
2.3	.9989	.9989	.9990	.9990	.9991	.9991	.9992	.9992	.9992	.9993
2.4	.9993	.9994	.9994	.9994	.9994	.9995	.9995	.9995	.9996	.9996

Areas and ordinates of *unit* normal curve in terms of x/σ

$$A = \frac{1}{\sigma\sqrt{2\pi}} \int_\mu^x e^{-(\mu-x)^2/2\sigma^2}\, dx \qquad O = \frac{1}{\sqrt{2\pi}} e^{-(\mu-x)^2/2\sigma^2}$$

Note that total area is 0.500, and A is fractional part between mean $(x = \bar{\mu})$ and x/σ. To find ordinate for specific problem, multiply O by N/σ.

x/σ	A	O	x/σ	A	O	x/σ	A	O
0.0	.0000	.3989	1.2	.3849	.1942	2.4	.4918	.0224
0.1	.0398	.3970	1.3	.4032	.1714	2.5	.4938	.0175
0.2	.0793	.3910	1.4	.4192	.1497	2.6	.4953	.0136
0.3	.1179	.3814	1.5	.4332	.1295	2.7	.4965	.0104
0.4	.1555	.3683	1.6	.4452	.1109	2.8	.4974	.0079
0.5	.1915	.3521	1.7	.4554	.0941	2.9	.4981	.0060
0.6	.2258	.3332	1.8	.4641	.0790	3.0	.4987	.0044
0.7	.2580	.3123	1.9	.4713	.0656	3.5	.49977	.00087
0.8	.2881	.2897	2.0	.4773	.0540	4.0	.49997	.00013
0.9	.3159	.2661	2.1	.4821	.0440	4.5	.49999	.00002
1.0	.3413	.2420	2.2	.4861	.0355	5.0	.49999	.00000
1.1	.3643	.2179	2.3	.4893	.0283			

6. COMMON INTEGRALS

In all of these indefinite integrals, a constant of integration should be added: e.g. $\int x\, dx = \frac{1}{2}x^2 + C$

1. $\int x^n\, dx = (n+1)^{-1}x^{n+1}$ (except for $n = -1$)

2. $\int x^{-1}\, dx = \ln x$

3. $\int (x^2 + a^2)^{-1}\, dx = \frac{1}{a}\tan^{-1}\frac{x}{a}$

4. $\int (x^2 - a^2)^{-1}\, dx = \frac{1}{2a}\ln\frac{x-a}{x+a}$

5. $\int e^x\, dx = e^x$

6. $\int a^x\, dx = (\ln a)^{-1}a^x$

7. $\int \sin x\, dx = -\cos x$

8. $\int \sin^2 x\, dx = \frac{1}{2}x - \frac{1}{4}\sin 2x = \frac{1}{2}(x - \sin x \cos x)$

9. $\int \sin^4 x\, dx = \frac{3}{8}x - \frac{1}{4}\sin 2x + \frac{1}{32}\sin 4x$

10. $\int \cos x\, dx = \sin x$

11. $\int \cos^2 x\, dx = \frac{1}{2}x + \frac{1}{4}\sin 2x = \frac{1}{2}(x + \sin x \cos x)$.

12. $\int \cos^4 x\, dx = \frac{3}{8}x + \frac{1}{4}\sin 2x + \frac{1}{32}\sin 4x$

13. $\int \tan x\, dx = -\ln \cos x$

14. $\int \cot x\, dx = \ln \sin x$

15. $\int \sec x\, dx = \ln(\sec x + \tan x) = \ln \tan\left(\frac{x}{2} + \frac{\pi}{4}\right)$

16. $\int \sec^2 x\, dx = \tan x$

17. $\int \csc x\, dx = \ln \tan\frac{x}{2} = \ln(\csc x - \cot x)$

18. $\int \csc^2 x\, dx = -\cot x$

19. $\int (a^2 - x^2)^{1/2}\, dx = \frac{x}{2}(a^2 - x^2)^{1/2} + \frac{a^2}{2}\sin^{-1}\frac{x}{a}$

20. $\int x^2(a^2 - x^2)^{1/2}\, dx = \frac{x}{8}(2x^2 - a^2)(a^2 - x^2)^{1/2} + \frac{a^4}{8}\sin^{-1}\frac{x}{a}$

21. $\int x^{-1}(a^2 - x^2)^{1/2}\, dx = (a^2 - x^2)^{1/2} + a\ln x^{-1}[a - (a^2 - x^2)^{1/2}]$

22. $\int (a^2 - x^2)^{-1/2}\, dx = \sin^{-1}\frac{x}{a}$

23. $\int x^2(a^2 - x^2)^{-1/2}\, dx = -\frac{x}{2}(a^2 - x^2)^{1/2} + \frac{a^2}{2}\sin^{-1}\frac{x}{a}$

24. $\int x^{-1}(a^2 - x^2)^{-1/2} dx = -\frac{1}{a} \ln x^{-1}[a + (a^2 - x^2)^{1/2}]$

25. $\int x^{-2}(a^2 - x^2)^{-1/2} dx = -(a^2x)^{-1}(a^2 - x^2)^{1/2}$

26. $\int (x^2 \pm a^2)^{1/2} dx = \frac{x}{2}(x^2 \pm a^2)^{1/2} \pm \frac{a^2}{2} \ln [x + (x^2 \pm a^2)^{1/2}]$

27. $\int x^2(x^2 \pm a^2)^{1/2} dx = \frac{x}{8}(2x^2 \pm a^2)(x^2 \pm a^2)^{1/2} - \frac{a^4}{8} \ln [x + (x^2 \pm a^2)^{1/2}]$

28. $\int x^{-1}(x^2 - a^2)^{1/2} dx = (x^2 - a^2)^{1/2} - a \sec^{-1} \frac{x}{a}$

29. $\int (x^2 \pm a^2)^{-1/2} dx = \ln [x + (x^2 \pm a^2)^{1/2}]$

30. $\int x^2(x^2 \pm a^2)^{-1/2} dx = \frac{x}{2}(x^2 \pm a^2)^{1/2} \mp \frac{a^2}{2} \ln [x + (x^2 \pm a^2)^{1/2}$

31. $\int x^{-1}(x^2 - a^2)^{-1/2} dx = \frac{1}{a} \sec^{-1} \frac{x}{a}$

32. $\int x^{-2}(x^2 \pm a^2)^{-1/2} dx = \mp(a^2x)^{-1}(x^2 \pm a^2)^{1/2}$

33. $\int x^{-1}(x^2 + a^2)^{1/2} = (x^2 + a^2)^{1/2} + a \ln x^{-1}[(x^2 + a^2)^{1/2} - a]$

34. $\int x^{-1}(x^2 + a^2)^{-1/2} = \frac{1}{a} \ln x^{-1}[(x^2 + a^2)^{1/2} - a]$

6. COMMON INTEGRALS

7. CHI SQUARE

Values of χ^2

n	$\alpha = 0.995$	$\alpha = 0.99$	$\alpha = 0.975$	$\alpha = 0.95$	$\alpha = 0.05$	$\alpha = 0.025$	$\alpha = 0.01$	$\alpha = 0.005$	v
1	0.0000393	0.000157	0.000982	0.00393	3.841	5.024	6.635	7.879	1
2	0.0100	0.0201	0.0506	0.103	5.991	7.378	9.210	10.597	2
3	0.0717	0.115	0.216	0.352	7.815	9.348	11.345	12.838	3
4	0.207	0.297	0.484	0.711	9.488	11.143	13.277	14.860	4
5	0.412	0.554	0.831	1.145	11.070	12.832	15.086	16.750	5
6	0.676	0.872	1.237	1.635	12.592	14.449	16.812	18.548	6
7	0.989	1.239	1.690	2.167	14.067	16.013	18.475	20.278	7
8	1.344	1.646	2.180	2.733	15.507	17.535	20.090	21.955	8
9	1.735	2.088	2.700	3.325	16.919	19.023	21.666	23.589	9
10	2.156	2.558	3.247	3.940	18.307	20.483	23.209	25.188	10
11	2.603	3.053	3.816	4.575	19.675	21.920	24.725	26.757	11
12	3.074	3.571	4.404	5.226	21.026	23.337	26.217	28.300	12
13	3.565	4.107	5.009	5.892	22.362	24.736	27.688	29.819	13
14	4.075	4.660	5.629	6.571	23.685	26.119	29.141	31.319	14
15	4.601	5.229	6.262	7.261	24.996	27.488	30.578	32.801	15
16	5.142	5.812	6.908	7.962	26.296	28.845	32.000	34.267	16
17	5.697	6.408	7.564	8.672	27.587	30.191	33.409	35.718	17
18	6.265	7.015	8.231	9.390	28.869	31.526	34.805	37.156	18
19	6.844	7.633	8.907	10.117	30.144	32.852	36.191	38.582	19
20	7.434	8.260	9.591	10.851	31.410	34.170	37.566	39.997	20
21	8.034	8.897	10.283	11.591	32.671	35.479	38.932	41.401	21
22	8.643	9.542	10.982	12.338	33.924	36.781	40.289	42.796	22
23	9.260	10.196	11.689	13.091	35.172	38.076	41.638	44.181	23
24	9.886	10.856	12.401	13.848	36.415	39.364	42.980	45.558	24
25	10.520	11.524	13.120	14.611	37.652	40.646	44.314	46.928	25
26	11.160	12.198	13.844	15.379	38.885	41.923	45.642	48.290	26
27	11.808	12.879	14.573	16.151	40.113	43.194	46.963	49.645	27
28	12.461	13.565	15.308	16.928	41.337	44.461	48.278	50.993	28
29	13.121	14.256	16.047	17.708	42.557	45.722	49.588	52.336	29
30	13.787	14.953	16.791	18.493	43.773	46.979	50.892	53.672	30

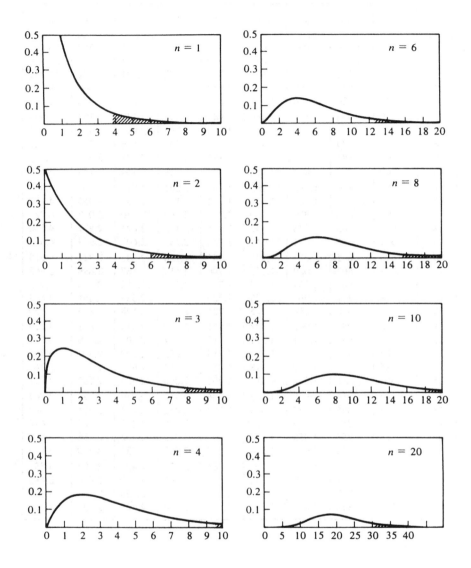

Chi square distributions for various values of n, the number of degrees of freedom. For n degrees of freedom, the sample size will usually be $n + 1$. The shaded regions correspond to probabilities of occurrence of less than 5% ($\alpha = 0.05$).

8. STUDENT'S *t* DISTRIBUTION

α n	.25	.20	.15	.10	.05	.025	.01	.005	.0005
1	1.000	1.376	1.963	3.078	6.314	12.706	31.821	63.657	636.619
2	.816	1.061	1.386	1.886	2.920	4.303	6.965	9.925	31.598
3	.765	.978	1.250	1.638	2.353	3.182	4.541	5.841	12.941
4	.741	.941	1.190	1.533	2.132	2.776	3.747	4.604	8.610
5	.727	.920	1.156	1.476	2.015	2.571	3.365	4.032	6.859
6	.718	.906	1.134	1.440	1.943	2.447	3.143	3.707	5.959
7	.711	.896	1.119	1.415	1.895	2.365	2.998	3.499	5.405
8	.706	.889	1.108	1.397	1.860	2.306	2.896	3.355	5.041
9	.703	.883	1.100	1.383	1.833	2.262	2.821	3.250	4.781
10	.700	.879	1.093	1.372	1.812	2.228	2.764	3.169	4.587
11	.697	.876	1.088	1.363	1.796	2.201	2.718	3.106	4.437
12	.695	.873	1.083	1.356	1.782	2.179	2.681	3.055	4.318
13	.694	.870	1.079	1.350	1.771	2.160	2.650	3.012	4.221
14	.692	.868	1.076	1.345	1.761	2.145	2.624	2.977	4.140
15	.691	.866	1.074	1.341	1.753	2.131	2.602	2.947	4.073
16	.690	.865	1.071	1.337	1.746	2.120	2.583	2.921	4.015
17	.689	.863	1.069	1.333	1.740	2.110	2.567	2.898	3.965
18	.688	.862	1.067	1.330	1.734	2.101	2.552	2.878	3.922
19	.688	.861	1.066	1.328	1.729	2.093	2.539	2.861	3.883
20	.687	.860	1.064	1.325	1.725	2.086	2.528	2.845	3.850
21	.686	.859	1.063	1.323	1.721	2.080	2.518	2.831	3.819
22	.686	.858	1.061	1.321	1.717	2.074	2.508	2.819	3.792
23	.685	.858	1.060	1.319	1.714	2.069	2.500	2.807	3.767
24	.685	.857	1.059	1.318	1.711	2.064	2.492	2.797	3.745
25	.684	.856	1.058	1.316	1.708	2.060	2.485	2.787	3.725
26	.684	.856	1.058	1.315	1.706	2.056	2.479	2.779	3.707
27	.684	.855	1.057	1.314	1.703	2.052	2.473	2.771	3.690
28	.683	.855	1.056	1.313	1.701	2.048	2.467	2.763	3.674
29	.683	.854	1.055	1.311	1.699	2.045	2.462	2.756	3.659
30	.683	.854	1.055	1.310	1.697	2.042	2.457	2.750	3.646
40	.681	.851	1.050	1.303	1.684	2.021	2.423	2.704	3.551
60	.679	.848	1.046	1.296	1.671	2.000	2.390	2.660	3.460
120	.677	.845	1.041	1.289	1.658	1.980	2.358	2.617	3.373
∞	.674	.842	1.036	1.282	1.645	1.960	2.326	2.576	3.291

Student's *t* Distribution

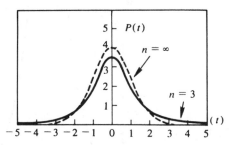

Comparison of Student's *t* distribution, where n = degrees of freedom. For $n = \infty$, $t = u$, and distribution is Gaussian with $\sigma = 1$.

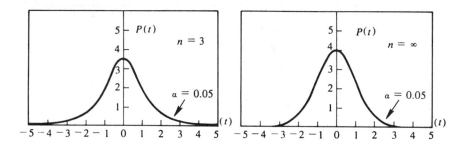

The shaded regions correspond to probability of 5% of *t* being greater than the indicated value. In the table of *t* distribution, the α value of probability is for $t > +t_a$. There is an equal probability for $t < -t_a$. Thus in the table on p. 95, for 3 degrees of freedom ($n = 3$), the 90% confidence level is for $t = 2.35$. 90% of *t* values for samples with $n = 3$ will fall between -2.35 and $+2.35$. There is a 5% probability that the *t* value will be greater than $+2.35$ ($\alpha = 0.05$).

8. STUDENT'S *t* DISTRIBUTION

9. HYPERBOLIC SINE, COSINE, AND TANGENT

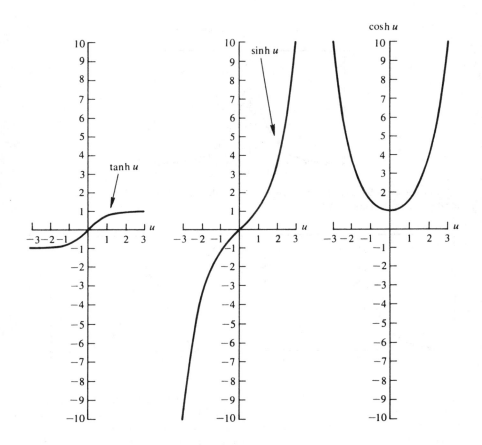

10. SERIES APPROXIMATIONS

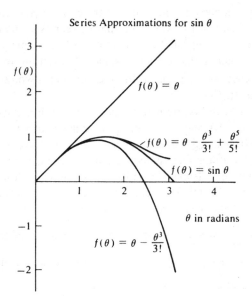

Series Approximations for $\sin \theta$

$f(\theta) = \theta$

$f(\theta) = \theta - \dfrac{\theta^3}{3!} + \dfrac{\theta^5}{5!}$

$f(\theta) = \sin \theta$

θ in radians

$f(\theta) = \theta - \dfrac{\theta^3}{3!}$

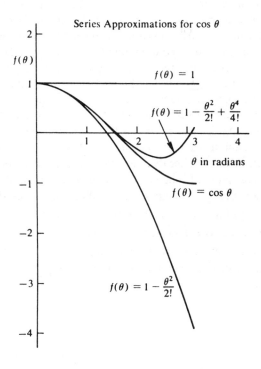

Series Approximations for $\cos \theta$

$f(\theta) = 1$

$f(\theta) = 1 - \dfrac{\theta^2}{2!} + \dfrac{\theta^4}{4!}$

θ in radians

$f(\theta) = \cos \theta$

$f(\theta) = 1 - \dfrac{\theta^2}{2!}$

11. PROPER CITATION OF ERROR LIMITS

How do we know what is meant when we read that a particular datum equals 10.0 ± 0.1 cm? Is this a personal guarantee of the experimenter that the results of all skillful measurements of this value will lie between 9.9 cm and 10.1 cm? Or does the writer mean that repeated measurements of the value yield a Gaussian distribution with a mean of 10.0 cm and a standard deviation (root mean square) of 0.1 cm? In that case, about $\frac{2}{3}$ of skillful measurements would produce values between 9.9 cm and 10.1 cm.

There is another possibility, frequently used in research reporting. The experimenter may have taken 100 measurements, plotted the distribution, and found that it had a mean of 10.0 cm and a standard deviation of *1 cm*. The experimenter would then be justified in reasoning that another run of 100 measurements would yield a similar distribution with a mean and standard deviation very close to the first one. According to the Central Limit Theorem (see p. 92), if the means of repeated runs were plotted, the *means* would have a Gaussian distribution with a standard deviation equal to $\dfrac{\sigma}{\sqrt{n}}$, where σ is the standard deviation of the original data distribution (in this case equal to 1 cm) and n is the number of measurements in one of the runs. Therefore the *standard deviation of the means* is equal to $\dfrac{1\ cm}{\sqrt{100}}$, or 0.1 cm.

But which of these possibilities is meant when the $\pm$ value is given? Unless the author specifies the meaning, or it is clearly implied by the experimental procedure, no one knows. How should *you* cite the error in reports? Define your error in words! If you are giving the maximum and minimum limits, say so, and explain how you chose those limits. Presumably the systematic errors (which do not produce a Gaussian distribution) were larger than the random errors. Perhaps you took only one or two measurements because there was no need for precision analysis. In that case you probably judged the error limits based on your experience and knowledge of the instruments and the procedure. If you did take enough measurements to yield a distribution curve, and if that curve is roughly bell-shaped, and if you judge that your systematic errors are smaller than the random error, then you are justified in using the standard deviation. If you cite the standard deviation of the distribution, and define it as such, then you are claiming that if one more measurement is made it stands a $\frac{2}{3}$ chance of having a value between $\bar{x} - \sigma$ *and* $\bar{x} + \sigma$. However, if you define σ to be the standard deviation of the mean ($\sigma_{mean} = \sigma_{distribution}/\sqrt{n}$), then you are claiming that if another set of n measurements is made, the mean of that set stands a $\frac{2}{3}$ chance of falling within the limits $\bar{x} \pm \sigma_{mean}$.

The standard deviation of a distribution is not necessarily related to the value of the mean or to the number of measurements in the sample. Note the distributions shown on p. 87, which have the same mean but different standard deviations. The standard I.Q. curve, for instance, has a mean of 100 and a standard deviation of 16 regardless of how many people are measured, providing that the sample is large enough to minimize statistical fluctuations. The binomial distribution, however, has a standard deviation of $\sqrt{np\,(1-p)}$ where n is the number of "throws," p is the probability of "success" in any one throw, and np is the most likely number of successes, usually assumed to be the number that is measured. (For instance, if you throw a die 36 times, and the probability of getting a 1 on any throw is $\frac{1}{6}$, the most likely number of 1's in 36 throws is 6.) Since p is usually small, $(1-p) \approx 1$, and $\sigma \approx \sqrt{np}$. For instance, if a Geiger counter measures 100 counts in one second, then the datum should be entered on the graph as $100 \pm \sqrt{100}$, or 100 ± 10. How does this work in the case of an I.Q. measurement? Suppose you test just one student and measure an I.Q. of 100. That tells you nothing about the standard distribution, but—assuming that the student's mistakes are random—suggests that $\frac{2}{3}$ of repeated measurements of the same student would yield test scores between 90 and 110. (Of course, it would be a major research project to construct a large number of truly duplicate exams in order to test the validity of such a conjecture.)

For binomial distributions, increasing the number of counts increases the absolute error which is the standard deviation, since $\sigma \approx \sqrt{np} = \sqrt{\text{number of counts}}$. For instance if your counter registers 100 counts, your datum is 100 ± 10. If you count 10 times as long and get 1000 counts, your absolute error is $\sqrt{1000}$ or approximately ± 30. The fractional error has decreased, however, from $100 \pm 10\%$ to $1000 \pm 3\%$. Would you do better to take 10 runs, each taking $\frac{1}{10}$ the time, getting about 100 counts each time? Then your error for each is ± 10. Combining 10 such runs, the combination error as given on p. 8 is: $\dfrac{\sigma}{\sqrt{n}} = \dfrac{10}{\sqrt{10}} = 3$. We gain nothing by dividing up the data runs, except perhaps a check on consistency of the apparatus.

11. PROPER CITATION OF ERROR LIMITS

K. $\dfrac{d^2x}{dt^2} + k\left(\dfrac{dx}{dt}\right)^2 = K$

This is the equation for acceleration equal to a constant minus a friction term proportional to the *square* of the velocity. It is the equation solved numerically in 15.2 for the specific case of an object falling in air. For almost all human-size objects moving with human-detectable speed in air or water, the Reynolds number is greater than 10, and the consequent drag is proportional to v^2. The solution for the speed involves a hyperbolic tangent, and provides an interesting comparison with the solution when the drag is proportional to v, the situation described by equation **G** (see p. 208).

For the case of an object falling in a fluid: $m_o \dfrac{dv}{dt} = (m_o - m_\rho)\, g - kv^2$, where m_o is the mass of the object, and m_ρ is the mass of the displaced fluid of density ρ. When $\dfrac{dv}{dt} = 0$, the resulting terminal velocity is $v_T = \sqrt{\dfrac{(m_o - m_\rho)\, g}{k}}$. In terms of v_T, the equation for v becomes: $\dfrac{dv}{dt} = \dfrac{(m_o - m_\rho)\, g}{m_o} - \dfrac{k}{m_o} v^2 =$

$$v_T^2 \frac{k}{m_o} - \frac{k}{m_o} v^2 = \frac{k}{m_o}(v_T^2 - v^2)$$

This equation can be transformed for integration:

$$\int \frac{dv}{(v_T^2 - v^2)} = \frac{k}{m_o} \int dt \Rightarrow \frac{1}{v_T} \tanh^{-1} \frac{v}{v_T} = \frac{k}{m_o} t \Rightarrow v = v_T \tanh\left(\frac{k}{m_o} v_T\right) t \Rightarrow$$

$$v = v_T \tanh\left(\frac{m_o - m_\rho}{m_o}\right) \frac{g}{v_T} t$$

The integration constant is zero if $v = 0$ at $t = 0$. To see the physical significance of this relationship, examine the hyperbolic tangent curve shown on page 254. For large t, the hyperbolic tangent equals 1 and so $v \to v_T$. When $t = \left(\dfrac{m_o}{m_o - m_\rho}\right)\dfrac{v_T}{g}$, $v = v_T \tanh 1 = 0.76\, v_T$. Note also that gt is the speed that the object would have in free fall at time t. Consequently, if the buoyant force is negligible, $v =$

$v_T \tanh\left(\dfrac{gt}{v_T}\right) = v_T \tanh\left(\dfrac{v_{\text{free fall at } t}}{v_T}\right)$. For gt small compared with v_T (at the beginning of the fall when v is small), $\tanh\left(\dfrac{gt}{v_T}\right) \approx \dfrac{gt}{v_T}$. Therefore $v \approx gt$ as we would expect, since the friction effect is still small.

INDEX

Statistics (*cont.*):
  central limit theorem, 92, 94
  chi square:
    definition and derivation, 98–101
    distribution curves, 251
    tables, 250
  confidence level, 94
  degrees of freedom, 89, 96
  Gaussian properties, 86–88
  line of regression, 27, 102
  mean, 86–88
  one-tail and two-tail tests, 97–98
  population, 85
  probable error, 88
  r coefficient, 101–106
  random variable, 85
  sample, 85, 89, 92–101
  standard deviation, 11, 87–89
    of sample, 89
  Student's t test, 96–98, 252, 253
  u test, 94–96
  variance, 88
Stellar magnitude scale, 71
Stirling's formula for factorial, 73
Student's t test 96–98, 252, 253

T

t test, in statistics, 96–98
Tangent (*see* Sinusoidal functions)
Tangent and perpendicular line equations, 124
Taylor's series, 194
Transforming bases:
  exponential, 56–57
  log, 66
Traveling wave function, 44, 166
Trigonometry:
  definitions of functions, 41
  identities and relationships, 50–52

U

u test in statistics, 94–96
Uncertainty in measurement, 1–15

Unit vectors:
  Cartesian, 155
  cylindrical, 142
  spherical, 144
Units, change of, 16–17

V

Variance, 88
Vectors, 153–161
  addition and subtraction, 153–154
  area represented by, 159
  commutation, 154, 157
  components, 154, 156
  cross product, 157–159
  cylindrical coordinates, 142
  definition, 153, 155
  determinant method for vector product, 158
  dot product, 156–157
  flux, 159–160
  multiplication, 156–159
  precession, 159
  pseudo-vectors, 155, 158
  right-hand rule, 157
  scalar product, 156–157
  scalars, 155
  spherical coordinates, 144–146
  unit vectors:
    Cartesian, 155
    cylindrical, 142
    spherical, 144
  vector product, 157–159
Velocity, 32, 45
Volume element:
  in cylindrical coordinates, 142
  in spherical coordinates, 144
Volumes of various figures, 151–152

W

Waves:
  complex number description, 165
  electromagnetic, derivation, 224–225
  function, 43
  traveling waves, 44, 166

8273